# The Kids' Guide to Birds of Virginia

## Fun Facts, Activities, and 87 Cool Birds

by Stan Tekiela

Adventure Publications

## DEDICATION

To all the children who enjoy the world of birds as much as I do.


## ACKNOWLEDGMENTS

Special thanks to the National Wildlife Refuge System, along with state and local agencies, both public and private, for stewarding the lands that are critical to the many bird species we so love.

Edited by Dan Downing, Sandy Livoti, and Brett Ortler
Cover and book design by Jonathan Norberg
Cover wavy border by ddok/Shutterstock.com
Illustrations by Elleyna Ruud
Range maps produced by Anthony Hertzel

**Cover photos:**
**Front:** Great Horned Owl by **DnDavis/Shutterstock.com;** Brown Pelican (main), American Kestrel, Eastern Bluebird, Wild Turkey, Yellow Warbler, and Wood Duck by **Stan Tekiela; Back:** Baltimore Oriole by **Stan Tekiela.**

**All other photos by Stan Tekiela** except:
pp. 122 (top inset), 172 (top inset) by **Dudley Edmondson**; p. 35 (inset) by **Brendan Klick**; p. 95 (inset) by **Maslowski Wildlife Productions**; pp. 40 (juvenile), 178 (in-flight juvenile) by **Brian K. Wheeler**; and pp. 94 (top inset), 174 (main) by **Jim Zipp**.

And these images used under license from Shutterstock.com:
**Cavan-Images:** 144 (main); **Chase D'animulls:** 112 (inset); **cliff collings:** 166 (inset); **Dennis Jacobsen:** 169 (inset); **Dr.Pixel:** 228; **Jill Schrock:** 200; **Melinda Fawver:** 144 (inset), 206 (inset); **Nickolay Stanev:** 38 (inset); **Paul Reeves Photography:** 104 (main), 160 (breeding inset); **Ray Hennessy:** 104 (top inset); **Steve Byland:** 148 (top inset); and **vagabond54:** 105 (inset).

To the best of the publisher's knowledge, all photos were of live birds. Some were photographed in a controlled condition.

10 9 8 7 6 5 4 3 2 1

**The Kids' Guide to Birds of Virginia: Fun Facts, Activities, and 87 Cool Birds**



Published by Adventure Publications
An imprint of AdventureKEEN
310 Garfield Street South
Cambridge, Minnesota 55008
(800) 678-7006
www.adventurepublications.net



Printed in China
Cataloging-in-Publication data is available from the Library of Congress
ISBN 978-1-64755-588-7 (pbk.); 978-1-64755-589-4 (ebook)

# Quick-Flip Color Guide

# TABLE OF CONTENTS

## Introduction



## The Birds



## Bird Food Fun for the Family

## COOL BIRDS IN VIRGINIA

*The Kids' Guide to Birds of Virginia* is a fun, easy-to-use guide for anyone interested in seeing and identifying birds. As a child, I enjoyed hours of watching birds come to a wooden feeder that my father built in our backyard. We were the only family in the neighborhood who fed birds, and we became known as the nature family.

Now, more people feed birds in their backyards than those who go hunting or fishing combined. Not only has it become very popular to feed and watch birds, but young and old alike are also identifying them and learning more about them.

The state of Virginia is a fantastic place to see all sorts of birds. In fact, more than 485 species are found here on a regular basis! That makes this one of the top places to watch an incredible variety of birds. In this field guide for Virginia, I'm featuring 87 of the most common of these great species.

We have marvelous habitats in Virginia that are perfect for birds. From our long coastline to mountain forests, Virginia has a marvelous mix of habitats and a wide variety of birds. Each habitat supports different kinds of birds. In the western portion of the state, there are tall Blue Ridge mountains with evergreen trees. This is a great place to see Pine Siskins. In the middle is a region called the Piedmont, which contains the foothills and coastal plain. This region is mainly old fields, streams, and woodlands. This is a good place to see Cardinals and House Wrens.

In addition to the mountains and Piedmont, Virginia has some very long coastlines. We have many different kinds of shorebirds such as the Ruddy Turnstone and several gull species that live here.

Virginia is located right on the migratory pathway of many small and large birds. Massive flocks of Tundra Swans pass through our states and winter along the coast, while birds such as the Ruby-throated Hummingbird arrive here from the warm tropics.

The weather here also plays a role in the kinds of birds we see. Yellow Warblers and Baltimore Orioles nest here during summer. Migrating shorebirds, such as the Snowy Egret, also come to Virginia to nest. On top of that, backyard birds, most notably Blue Jays and American Goldfinches, enjoy our seasons year-round.

As you can see, Virginia is a terrific place to watch all kinds of cool birds. It is my sincere hope that you and your family will like watching and feeding birds as much as I did with my family when I was a kid. Let this handy book guide you into a lifetime of appreciating birds and nature.

## BODY BASICS OF A BIRD

It's good to know the names of a bird's body parts. The right terminology will help you describe and identify a bird when you talk about it with your friends and family.

The basic parts of a bird are labeled in the illustration below. This drawing is a combination (composite) of several birds and should not be regarded as one particular species.

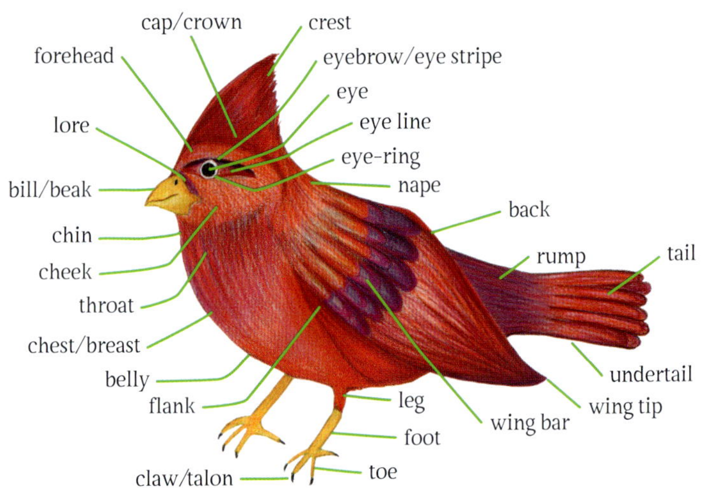


## AMAZING NESTS

I am fascinated with bird nests! They are amazing structures that do more than just provide a place for egg laying. Nests create a small climate-controlled environment that's beneficial for both keeping the eggs warm and raising the young after they hatch.

From the high treetops to the ground, there are many kinds of nests. Some are simple, while others are complex. In any case, they function in nearly the same way. Nests help to contain the eggs so they don't roll away. They also help to keep baby birds warm on cold nights, cool on hot days, and dry during rain.

The following illustrations show the major types of nests that birds build in Virginia.

A **ground nest** can be a mound of plant materials on the ground or in the water. Some are just a shallow spot scraped in the earth.

A **platform nest** is a cluster of sticks with a depression in the center. It is secured to the platform of a tree fork or to several tree branches.

A **cup nest** has a cupped interior, like a bowl.

A **pendulous nest** is a woven nest that hangs and swings freely, like a pendulum, from a branch.

A **cavity nest** is simply a cavity or hole, usually in a tree.

The first step in nest building is to choose an appropriate site. Each bird species has a unique requirement for this. Some birds, such as American Robins, just need a tree branch. Others, like Eastern Bluebirds, look for a cavity and build their nest inside. Still others, such as Killdeer, search for camouflaged ground to scrape out a nest. Sometimes birds such as Turkey Vultures don't bother building a nest at all if they spot a hard-to-reach cliff or rocky ledge where it will be safe to lay their eggs.

Nest materials usually consist of common natural items found in the area, like sticks or dried grass. Birds use other materials, such as mud or spiderwebs, to glue the materials together.

One of the amazing things about nest construction is that the parents don't need building plans or tool belts. They already know by instinct how to build nests, and they use their beaks and feet as their main tools.

To bring in nesting materials, birds must make many trips back and forth to the nest site. Most use their beaks to hold as much material as possible during each trip. Some of the bigger birds, like Bald Eagles, use larger materials, such as thick sticks and thin branches. They grasp and carry these items with their feet.

Nest building can take two to four days or longer, depending on the species and nest type. The simpler the nest, the faster the construction. Mourning Dove parents, for example, take just a few days to collect one to two dozen sticks for their platform nest. Woodpecker pairs, however, work for upwards of a week to **excavate**, or dig out, a suitable nesting cavity. Large and more complicated platform nests, such as a Bald Eagle nest, may take weeks or even a month to complete, but these can be used for years and are worth the extra effort.

## WHO BUILT THAT NEST?

In the majority of bird species, the chief builder is the female. In other species, both the female and the male typically share in the construction equally.

In general, when male and female birds look vastly different, the female does most of the work. When the male and female look alike or appear very similar, they tend to share the tasks of nest-building and feeding the young. Alternatively, some species of woodpeckers have a different building plan. When they chisel out a nesting chamber, often the male does more of the work after the female has chosen the site.

## ATTRACTING BIRDS WITH FEEDERS

To get more birds to visit your yard, an easy way to invite them is to put out bird feeders. Bird feeders are often as unique as the birds themselves, so the types of feeders you use really depend on the kinds of birds you're trying to attract.

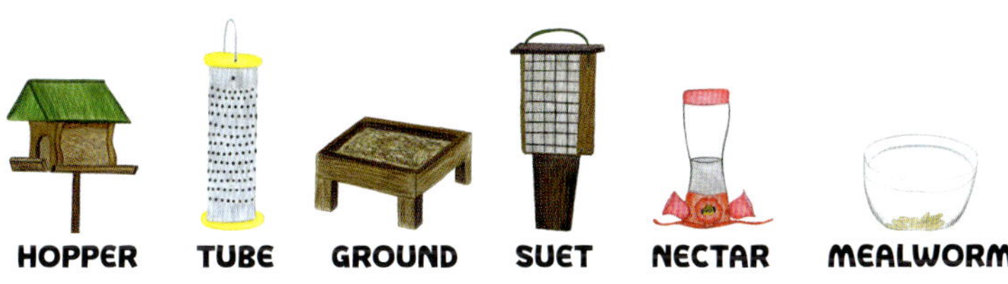


**Hopper feeders** are often wooden or plastic. Designed to hold a large amount of seeds, they often have a slender opening along the bottom, which dispenses the seeds. Birds land along the sides and help themselves to the food. Hopper feeders work well as main feeders in conjunction with other types of feeders. They are perfect for offering several kinds of seed mixes for cardinals, finches, nuthatches, chickadees, and more.

**Tube feeders** with large seed ports and multiple perches are very popular. Often mostly plastic, they tend to be rugged enough to last several years and can be easily cleaned. These feeders are great for black oil sunflower seeds and seed mixes, which are favorites of cardinals and all the other bird species that also visit hopper feeders.

Some tube feeders have small holes, allowing incredibly tiny thistle seeds to be dispensed just a few at a time. Use this kind of feeder to offer Nyjer seed, which will attract various finches.

Other styles of tube feeders have a wire mesh covering with openings large enough for birds to extract one of their favorite foods—peanuts out of the shell. Most birds enjoy peanuts, so these feeders will be some of the most popular in your yard. Another variety of tube feeder has openings large enough for peanuts in the shell. These are also very popular with the birds.

**Ground feeders** allow a wide variety of birds to access the food. The simplest and easiest feeders to use, they consist of a flat platform with a lip around the edges to keep seeds and corn from spilling out. Some have a roof to keep rain and snow off the food. With or without a roof, drainage holes in the bottom are important. Ground feeders will bring in towhees and many other birds to your backyard, including doves and even mallards if you're near water.

**Suet feeders** are simply wire cages that hold cakes of **suet**. The wire allows woodpeckers, nuthatches, and other birds to cling securely to the feeder while pecking out chunks of suet. The best suet feeders have a vertical extension at the bottom where a woodpecker can brace its tail and support itself while feeding. These are called tail-prop suet feeders.

**Nectar feeders** are glass or plastic containers that hold sugar water. These feeders usually have plastic parts that are bright red, a color that is extremely attractive to hummingbirds, but orioles and woodpeckers will also stop for a drink. They often have up to four ports for access to the liquid and yellow bee guards to prevent bees from getting inside.

**Mealworm feeders** can be very basic—a simple glass or plastic cup or container will do. Pick one with sides tall enough and make sure the material is slippery enough to stop the lively mealworms from crawling out. Bluebirds especially love this wiggly treat!

## HOW TO USE THIS GUIDE

Birds move pretty fast, so you don't often get a lot of time to observe them. To help you quickly find the birds in the book, this guide is organized by color. Simply note the most prominent color of the bird you've seen. A male Hairy Woodpecker, for example, is black and white and has a red mark on its head. Since this bird is mostly black and white, you would find it in the black-and-white section.

Within each color section, the birds are organized by size, from small to large. Use the Real Quick sidebar to find the size that your bird appears to be.

When the male and female of a species are different colors (like the Wood Duck pair below), they are shown in their own color sections. In these cases, the opposite sex is included in an inset photo with a page reference so you can easily turn to it.

If you already know the name of the bird you've seen, use the Checklist/Index to get the page number, and flip to it to learn more about the bird.

To further help you with identification, check the range maps to see where and when the bird you have sighted is normally found in Virginia. Range maps capture our current knowledge of where the birds are during a given year (presence) but do not indicate how many birds are in the area (density). In addition, since birds fly around freely, it's possible to see them outside of their ranges. So please use the maps to get a general idea of where the birds are most likely to be seen.

For more about the information given for each bird in this guide, turn to the Northern Cardinal sample on pgs. 16–17.

While you're learning about birds and identifying them, don't forget to check out the fun-filled things to do starting on pg. 220. Score a big hit with the birds in your yard by creating tasty treats or making your own bird food from the recipes. Put out some nesting materials to help birds build their nests. Consider signing up for a cool citizen science project suitable for the entire family. These are just a few of the activities that are such great fun you'll want to do them all!

# Northern Cardinal

Common name

**Look for the black mask**

Field markings that help identify the bird

**MALE**



Colored border shows the color section of the opposite sex

Turn to the page number to see the opposite sex of the species

**FEMALE**
pg. 107

Color tab

**Mostly Red**

**REAL QUICK**

**What to look for:**

outstanding features; may include other plumages and descriptions

**Where you'll find them:**

where you're most likely to see the bird

**Calls and songs:**

songs, calls, and other sounds the bird makes

**On the move:**

anything about flight, flocks, travel, and other movements

**What they eat:**

foods the bird eats and the kinds of feeders it visits

**Nest:**

type of nest; may include nest site, materials, and more

**Eggs, chicks, and childcare:**

number of eggs, color and marks; **incubation** and feeding duties; may include how many broods

**Spends the winter:**

where the bird goes when it's cold or when food is scarce

Length from head to tail

Size
8–9"

Type of nest the bird calls home

Nest
CUP

Type of feeder the bird generally visits

Feeder
HOPPER

Range map

year-round
summer
migration
winter

The bold word means it is defined in the glossary

After you've seen it, checkmark it

SAW IT!

## STAN'S COOL STUFF

Fun and interesting facts about the bird. Information not typically found in other field guides.

# Eastern Towhee

**Look for the black head**

**What to look for:**
mostly black bird with rusty sides, a white belly, red eyes, and a long black tail with a white tip

**Where you'll find them:**
shrubby areas with short trees and thick bushes, backyards and parks

**Calls and songs:**
calls "tow-hee" distinctly; also has a characteristic **call** that sounds like "drink-your-tea"

**On the move:**
short flights between shrubby areas and heavy **cover**; flashes white wing patches during flight

**What they eat:**
insects, seeds and fruit; comes to ground feeders

**Nest:**
cup; Mom constructs the nest

**Eggs, chicks, and childcare:**
3–4 creamy-white eggs with brown marks; Mom incubates the eggs; Dad and Mom feed the young

**Spends the winter:**
in southern states and South America; some stay in Virginia

**REAL QUICK**

Size
7–8"

Nest
CUP

Feeder
GROUND

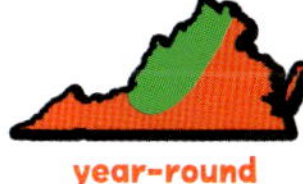

year-round
summer

SAW IT!

**STAN'S COOL STUFF**

The towhee is named for its distinctive "tow-hee" call. It hops backward with both feet, raking leaves to find insects and seeds. In southern coastal states, some have white eyes, and others have red eyes.

# Brown-headed Cowbird

**Look for the brown head**

MALE

FEMALE
pg. 99

**What to look for:**
glossy black bird with a chocolate-brown head and a sharp, pointed gray bill

**Where you'll find them:**
forest edges, open fields, farmlands, and backyards

**Calls and songs:**
sings a low, gurgling song that sounds like water moving; cowbird young are raised by other bird parents, but they still end up singing and calling like their own parents, whom they've never heard

**On the move:**
Mom flies quietly to another bird's nest, swiftly lays an egg, then flies quickly away

**What they eat:**
insects and seeds; visits seed feeders

**Nest:**
doesn't nest; lays eggs in the nests of other birds

**Eggs, chicks, and childcare:**
5–7 white eggs with brown marks; the **host** bird incubates any number of cowbird eggs in her nest and feeds the cowbird young along with her own

**Spends the winter:**
in Virginia; moves around to find food

**REAL QUICK**

Size
**7½"**

Nest
**NONE**

Feeder
**TUBE OR HOPPER**

year-round

SAW IT!

**STAN'S COOL STUFF**

Cowbirds are **brood parasites**, meaning they don't nest or raise their own families. Instead, they lay their eggs in other birds' nests, leaving the host birds to raise their young. Cowbirds have laid their eggs in the nests of more than 200 other bird species.

# European Starling

**Look for the glittering, iridescent feathers**

**What to look for:**
shiny and **iridescent** purplish black in spring and summer, speckled in fall and winter; yellow bill in spring, gray in fall; pointed wings and a short tail

**Where you'll find them:**
lines up with other starlings on power lines; found in all habitats but usually associated with people, farms, suburban yards, and cities

**Calls and songs:**
mimics the songs of up to 20 bird species; mimics other sounds, even imitating the human voice

**On the move:**
large family groups gather with blackbirds in fall

**What they eat:**
bugs, seeds, and fruit; visits seed and **suet** feeders

**Nest:**
cavity, filled with dried grass; often takes a cavity from other birds

**Eggs, chicks, and childcare:**
4–6 bluish eggs with brown marks; Mom and Dad sit on the eggs and feed the babies

**Spends the winter:**
in Virginia; moves around to find food

**REAL QUICK**

Size
7½"

Nest
CAVITY

Feeder
TUBE OR HOPPER

year-round

SAW IT!

**STAN'S COOL STUFF**

The starling is a mimic that can sound like any other bird. It's not a native bird; 100 starlings from Europe were introduced to New York City in 1890–91. Today, European Starlings are one of the most numerous songbirds in the country.

# Red-winged Blackbird

**Look for the red-and-yellow shoulder patches**

**What to look for:**
black bird with red-and-yellow shoulder patches on upper wings; shoulder patches can be partially or completely covered up

**Where you'll find them:**
around marshes, wetlands, lakes, and rivers

**Calls and songs:**
male sings and repeats calls from cattail tops and the surrounding **vegetation**

**On the move:**
flocks with as many as 10,000 birds gather in autumn, often with other blackbirds

**What they eat:**
seeds in spring and autumn, insects in summer; visits seed and **suet** feeders

**Nest:**
cup in a thick stand of cattails over shallow water

**Eggs, chicks, and childcare:**
3–4 speckled bluish-green eggs; Mom does all the incubating, but both parents feed the babies

**Spends the winter:**
in Virginia; moves around in winter to find food

**REAL QUICK**

Size
8½"

Nest
CUP

Feeder
TUBE OR HOPPER

year-round

SAW IT!

**STAN'S COOL STUFF**

During autumn and winter, thousands of these birds gather in farm fields, wetlands, and marshes. Come spring, males sing to defend territories and show off their wing patches (**epaulets**) to the females. Later, males can be aggressive when defending their nests.

# Common Grackle

Look for the shiny bluish-black head

**What to look for:**
shiny bluish-black **iridescent** head, a purplish-brown body, and super-bright golden eyes

**Where you'll find them:**
evergreen trees and shrubs, suburban and urban yards, open fields

**Calls and songs:**
gives a loud, raspy **call**

**On the move:**
travels in large flocks with other blackbirds; flight is usually level as opposed to an up-and-down pattern; male holds his tail in a deep V shape

**What they eat:**
fruit, seeds, and bugs; visits seed and **suet** feeders

**Nest:**
cup, usually in a **colony** of up to 75 mated pairs

**Eggs, chicks, and childcare:**
4–5 speckled greenish-white eggs; Mom sits on the eggs; Mom and Dad give food to the babies

**Spends the winter:**
in Virginia

**REAL QUICK**

Size
**11–13"**

Nest
**CUP**

Feeder
**HOPPER**

year-round

**SAW IT!**

**STAN'S COOL STUFF**

The Common Grackle is a member of the blackbird family. Compared to most birds, it has strong muscles to open its mouth. The muscles help it to pry apart crevices, where it finds bugs to eat. It's kind of like playing hide-and-seek for its food.

# Common Gallinule

Look for the red bill with a yellow tip

**What to look for:**
nearly black overall with a yellow-tipped red bill, a red forehead, and yellowish-green legs

**Where you'll find them:**
freshwater marshes and lakes

**Calls and songs:**
gives a series of fast, high-pitched clucks

**On the move:**
walks on floating **vegetation** or swims while on the hunt for bugs

**What they eat:**
insects, snails, seeds, green leaves, fruit, and roots

**Nest:**
ground nest; Mom and Dad build it with cattails and other wetland plants

**Eggs, chicks, and childcare:**
2–10 brown eggs with dark marks; Mom and Dad take turns incubating; young usually leave the nest within a few hours after hatching, but they stay with their family for a few months

**Spends the winter:**
in Florida, Central America and South America; some winter in eastern Virginia

**REAL QUICK**

Size
**13–15"**

Nest
**GROUND**

Feeder
**NONE**

year-round
summer

**SAW IT!**

**STAN'S COOL STUFF**

This duck-like bird is also called the Pond Chicken. It was once known as the Common Moorhen. Females are known to lay some of their eggs in other Common Gallinule nests. In the water, the young ride around on the backs of the adults.

# American Coot

Look for the white bill

**What to look for:**
gray-to-black bird with a duck-like white bill, red eyes

**Where you'll find them:**
in large flocks on open water

**Calls and songs:**
a unique series of creaks, groans, and clicks

**On the move:**
bobs head while swimming; takes off from water by scrambling across it with wings flapping; huge flocks of up to 1,000 birds gather for migration; migrates at night

**What they eat:**
insects and aquatic plants

**Nest:**
ground nest floating in water, anchored to plants

**Eggs, chicks, and childcare:**
9–12 speckled pinkish-tan eggs; Mom and Dad sit on the eggs and feed the **hatchlings**

**Spends the winter:**
in Virginia and southern states

**REAL QUICK**

Size
13–16"

Nest
GROUND

Feeder
NONE

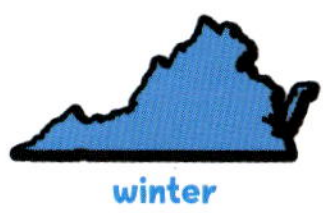

winter

SAW IT!

**STAN'S COOL STUFF**

The coot is not a duck. Instead of webbed feet, it has large lobed toes! It's smaller than most other **waterfowl**, and it is a great diver and swimmer. You won't see it flying, but you may spot one trying to escape from a Bald Eagle (pg. 65). It's also called Mud Hen.

# Boat-tailed Grackle

**Look for the very long tail**

pg. 121

**What to look for:**
glossy (**iridescent**) bluish-black bird with a very long tail and bright-yellow eyes

**Where you'll find them:**
coastal saltwater marshes and inland marshes

**Calls and songs:**
noisy, giving several harsh, high-pitched calls and several squeaks

**On the move:**
travels in large flocks with other blackbirds; flight is typically level, not in an up-and-down pattern

**What they eat:**
insects, berries, seeds, grains, and fish; comes to seed and **suet** feeders

**Nest:**
cup; Mom makes it with mud or cow dung and grass; nests twice each year in a small **colony**

**Eggs, chicks, and childcare:**
2–4 pale greenish-blue eggs with brown marks; Mom incubates the eggs and feeds the babies

**Spends the winter:**
doesn't **migrate**; stays in coastal Virginia year-round and moves around to find food

**REAL QUICK**

Size
**15–17"**

Nest
**CUP**

Feeder
**HOPPER**

year-round

**SAW IT!**

**STAN'S COOL STUFF**

Boat-tails are sometimes seen picking bugs off the backs of cattle. Nests in cattail stands. Forms large flocks in winter. Got its name by the way the male holds its tail in flight, forming a V like the keel of a boat.

Look for the small head

**What to look for:**
glossy black bird with a small head and black bill, a long black tail, and black feet

**Where you'll find them:**
all habitats—coastlines, rivers, lakes, wetlands, wilderness, rural, suburban, cities

**Calls and songs:**
gives a high-pitched, nasal "cah" **call**; imitates other birds and people

**On the move:**
gathers in winter flocks of up to 100 birds

**What they eat:**
insects, dead carcasses (**carrion**), clams, mussels, snails, slugs and other **mollusks**, berries and seeds; comes to seed and **suet** feeders

**Nest:**
platform, often builds a stick nest in a palm tree; nests in a small **colony**

**Eggs, chicks, and childcare:**
4–5 speckled blue or grayish-green eggs; Mom and Dad sit on the eggs and feed the young

**Spends the winter:**
in Virginia

**REAL QUICK**

Size
16"

Nest
PLATFORM

Feeder
HOPPER

year-round
summer

SAW IT!

**STAN'S COOL STUFF**

This crow is mainly a bird of the coast and major rivers, but it's all over Virginia. It breaks open mollusk shells by dropping them onto rocks from above. You can identify it by its call. It is higher in pitch than the call of the American Crow (pg. 37).

# American Crow

**What to look for:**
glossy black all over and a black bill

**Where you'll find them:**
all habitats—wilderness, rural, suburban, cities

**Calls and songs:**
a harsh "caw" **call**; imitates other birds and people

**On the move:**
flaps constantly and glides downward; moves around to find food; gathers in huge communal flocks of more than 10,000 birds during winter

**What they eat:**
fruit, insects, mammals, fish, and dead carcasses (**carrion**); visits seed and **suet** feeders

**Nest:**
platform; adds bright or shiny items and often uses the same site every year if a Great Horned Owl (pg. 139) hasn't taken it

**Eggs, chicks, and childcare:**
4–6 speckled bluish-to-olive eggs; Mom sits on the eggs; Mom and Dad feed the youngsters

**Spends the winter:**
in Virginia; moves around in winter to find food, often to city interiors

**REAL QUICK**

Size
18"

Nest
PLATFORM

Feeder
HOPPER

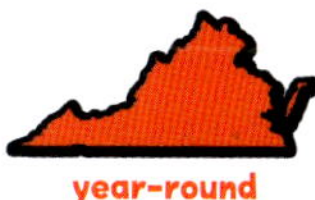

year-round

SAW IT!

**STAN'S COOL STUFF**

The crow is one of the smartest of all birds. It's very social and often entertains itself by chasing other birds. It eats roadkill but avoids being hit by vehicles. Some can live as long as 20 years! Crows without mates, called helpers, help to raise the young.

# Black Vulture

**Look for the naked dark gray head**

## Mostly Black

**What to look for:**
naked dark-gray head and legs, an ivory bill and a short tail; appears black in flight with light-gray wing tips

**Where you'll find them:**
in trees, sunning itself with wings outstretched, drying after a rain

**Calls and songs:**
mostly **mute**, just grunts and groans

**On the move:**
holds wings straight out to the sides during flight

**What they eat:**
dead carcasses (**carrion**); may capture small live mammals; parents **regurgitate** food for the young

**Nest:**
no nest, or on a stump or the ground; may use an empty nest; often nests with other Black Vultures

**Eggs, chicks, and childcare:**
1–3 light-green eggs with dark marks; Mom and Dad do the **incubation** and feed the babies

**Spends the winter:**
doesn't **migrate**; most stay in Virginia year-round

**REAL QUICK**

Size
**25–28"**

Nest
**NONE**

Feeder
**NONE**

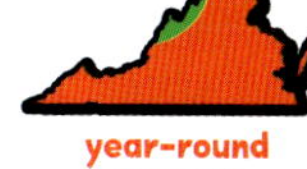

year-round
summer

SAW IT!

**STAN'S COOL STUFF**

People also call this bird the Black Buzzard. It's not as good at finding carrion as the Turkey Vulture (pg. 41), so its sense of smell may be weaker. If startled, especially at the nest, it regurgitates with power and accuracy. Families stay together for up to a year.

# Turkey Vulture

**Look for the naked red head**

**What to look for:**
naked red head and legs and an ivory bill; juvenile has a gray-to-blackish head and bill

**Where you'll find them:**
in trees, sunning itself with wings outstretched, drying after a rain

**Calls and songs:**
mostly **mute**, just grunts and groans

**On the move:**
holds wings in an upright V in flight, teetering from wing tip to wing tip as it soars and hovers

**What they eat:**
dead carcasses (**carrion**); parents **regurgitate** food for their young

**Nest:**
no nest, or in a minimal nest on a cliff, in a cave, or even sometimes in a hollow tree trunk

**Eggs, chicks, and childcare:**
1–3 white eggs with brown marks; Mom and Dad incubate the eggs and feed the baby vultures

**Spends the winter:**
in Virginia; moves around to find food

**REAL QUICK**

Size
**26–32"**

Nest
**NONE**

Feeder
**NONE**

year-round

**SAW IT!**

**STAN'S COOL STUFF**

This is one of the few birds with a good sense of smell. It has a strong bill for tearing apart flesh. Unlike hawks and eagles, it has weak feet more suited for walking than grasping wiggly **prey**. The bare head reduces its risk of getting diseases from carcasses.

# Double-crested Cormorant

**Look for the large, hooked bill**

**What to look for:**
large black waterbird with unusual blue eyes, a long snake-like neck, and large gray bill with a yellow base and hooked tip

**Where you'll find them:**
usually roosts in large groups in trees near water

**Calls and songs:**
grunts, pops, and groans—none are pleasant sounds at all!

**On the move:**
swims underwater to catch fish, holding its wings at its sides; flies in a large V-shaped formation

**What they eat:**
small fish and aquatic insects

**Nest:**
platform near or over open water, in a **colony**

**Eggs, chicks, and childcare:**
3–4 bluish-white eggs; parents take turns sitting on the eggs and feeding the young

**Spends the winter:**
some stay in coastal Virginia; southern states, Mexico, and Central America

**REAL QUICK**

Size
**31-35"**

Nest
**PLATFORM**

Feeder
**NONE**

year-round
migration

SAW IT!

**STAN'S COOL STUFF**

This bird's outer feathers are different from its inner ones; the outer feathers soak up water, but its body feathers don't. It opens its wings and uses the sun and wind to dry out. "Double-crested" refers to the two unusual crests on its head, but these aren't often seen.

# Downy Woodpecker

**Look for the small, short bill**

**What to look for:**
spotted wings, white belly, red mark on the back of the head, and a small, short bill; female lacks a red mark on the head

**Where you'll find them:**
wherever trees are present

**Calls and songs:**
repeats a high-pitched "peek-peek" **call**; drums on trees or logs with its bill to announce its territory

**On the move:**
flies in an up-and-down pattern; makes short flights from tree to tree

**What they eat:**
insects and seeds; visits **suet** and seed feeders

**Nest:**
cavity in a dead tree; digs out a perfectly round entrance hole; the bottom of the cavity is wider than the top, and it's lined with fallen woodchips

**Eggs, chicks, and childcare:**
3–5 white eggs; Mom incubates the eggs; both parents take care of the kiddies

**Spends the winter:**
in Virginia; moves around to find food

**REAL QUICK**

Size
6"

Nest
CAVITY

Feeder
SUET

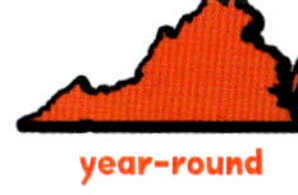

year-round

SAW IT!

**STAN'S COOL STUFF**

The Downy is abundant and widespread where trees are present. Like other woodpeckers, it pulls insects from tiny places with its long, barbed tongue. It has stiff tail feathers, which help to support it as it clings to trees. During winter, it will roost in a cavity.

# Rose-breasted Grosbeak

**Look for the rosy-red breast patch**

pg. 97

**What to look for:**
black-and-white bird with a large rosy-red patch in the center of the breast and a large ivory bill

**Where you'll find them:**
in the woods, usually in pairs around seed feeders

**Calls and songs:**
male and female sing a rich, robin-like song with a distinctive chip **note** sprinkled in the tune; male is much louder and clearer than the female and sings to her from high up in trees

**On the move:**
short flights from tree to tree with rapid wingbeats; male flashes white wing patches in flight

**What they eat:**
insects, seeds, and fruit; visits seed feeders

**Nest:**
cup in a mature **deciduous** forest, in a small tree

**Eggs, chicks, and childcare:**
3–5 speckled blue-green eggs; Mom and Dad sit on the eggs and feed the young chicks; juveniles come with the adults to bird feeders

**Spends the winter:**
in Mexico, Central and South America

**REAL QUICK**

Size
7-8"

Nest
CUP

Feeder
TUBE OR HOPPER

summer
migration

**STAN'S COOL STUFF**

The rose breast patch varies in size and shape among the males. "Grosbeak" refers to the large bill which the bird uses to crush seeds. In spring, males return before the females. Males become territorial when females arrive and reduce their visits to feeders.

# Yellow-bellied Sapsucker

**Look for the red chin**

**What to look for:**
checkered back, yellow chest and belly, and a red forehead, crown, and chin; female has a white chin

**Where you'll find them:**
small woods, forests, suburban, and rural areas, small to medium trees that have rows of sap holes

**Calls and songs:**
quiet with few vocalizations but will meow like a cat; drums on hollow tree branches irregularly (not in a regular pattern, like other woodpeckers)

**On the move:**
short up-and-down flights with rapid wingbeats

**What they eat:**
insects, nutritious sap in trees; visits **suet** feeders

**Nest:**
cavity that Mom and Dad **excavate**, often in a live tree—but a dead tree will work, too!

**Eggs, chicks, and childcare:**
5–6 white eggs; Mom and Dad incubate the eggs and feed the **brood**

**Spends the winter:**
most stay in Virginia; southern states, Mexico, and Central America

**REAL QUICK**

Size
**8–9"**

Nest
**CAVITY**

Feeder
**SUET**

summer
migration
winter

**SAW IT!**

**STAN'S COOL STUFF**

Sapsuckers are woodpeckers that drill rows of holes in trees to get to the sap. **Tree sap** is full of minerals and other nutrients. Sapsuckers don't actually suck the sap out of the holes; rather, they lap it with their long, bristly tongues. Oozing sap also attracts bugs to eat.

# Hairy Woodpecker

Look for the large bill

**What to look for:**
spotted wings, white belly, large bill, and red mark on the back of the head; female lacks a red mark

**Where you'll find them:**
forests and wooded backyards, parks

**Calls and songs:**
a sharp chirp before landing on feeders; drums on hollow logs, branches, or stovepipes in spring

**On the move:**
short up-and-down flights from tree to tree with rapid wingbeats

**What they eat:**
insects, nuts, seeds; visits **suet** and seed feeders

**Nest:**
cavity; prefers a live tree; excavates a larger, more oval entry than the round hole of the Downy Woodpecker (pg. 45); usually excavates under a branch, which helps to shield the entrance

**Eggs, chicks, and childcare:**
3–6 white eggs; parents sit on the eggs and bring food to feed their babies

**Spends the winter:**
in Virginia

**REAL QUICK**

Size
9"

Nest
CAVITY

Feeder
SUET

year-round

SAW IT!

**STAN'S COOL STUFF**

The Hairy is nearly identical to the Downy Woodpecker, but it's larger and has a larger, longer bill. It has a barbed tongue, which it uses to pull out bugs from trees. At the base of its bill, tiny bristle-like feathers protect its nostrils from excavated wood dust.

# Red-bellied Woodpecker

**Look for the black-and-white striped back**

**What to look for:**
zebra-striped back, white rump, red crown, and red nape of neck; female has a light-gray crown

**Where you'll find them:**
shady woodlands, forest edges, and backyards

**Calls and songs:**
calls a loud "querrr" and a low "chug-chug-chug"

**On the move:**
rapid wingbeats in flight, going up and down like a roller coaster

**What they eat:**
spiders and centipedes as well as beetles and other insects; nuts and fruit; visits **suet** and seed feeders

**Nest:**
cavity in a dead tree; excavates a new hole in last year's tree below the previous cavity

**Eggs, chicks, and childcare:**
4–5 white eggs; Mom incubates the eggs during the day and Dad takes night duty; both parents feed the baby woodpeckers

**Spends the winter:**
in Virginia; moves around to find food

**REAL QUICK**

Size
**9–9½"**

Nest
**CAVITY**

Feeder
**SUET**

year-round

**SAW IT!**

**STAN'S COOL STUFF**

This bird is named for its faint pink belly patch. It excavates dead wood to find bugs to eat and hammers acorns and berries into cracks in trees to store for winter food. The population and range are increasing across the country.

# Ruddy Turnstone

**Look for the black-and-white head**

**What to look for:**
unusual black-and-white head pattern, black bib, white throat and belly, black-and-rust wings and back, slightly upturned black bill; winter **plumage** has a brown-and-white head and chest pattern

**Where you'll find them:**
rocky beaches, sandy beaches, crabbing operations (where it eats scraps from nets)

**Calls and songs:**
if stressed, gives a fast, high-pitched alarm **call**

**On the move:**
turns over stones on rocky beaches to find food

**What they eat:**
aquatic insects, fish, snails and other **mollusks**, crabs and other **crustaceans**, worms, bird eggs

**Nest:**
ground nest; just Mom builds it

**Eggs, chicks, and childcare:**
3–4 speckled olive-colored eggs; Dad and Mom sit on the eggs; Mom leaves before the babies leave the nest (**fledge**), so Dad feeds the babies by himself

**Spends the winter:**
in coastal Virginia; other southern coastal states, Mexico, Central America, and South America

**REAL QUICK**

Size
9½"

Nest
GROUND

Feeder
NONE

migration
winter

SAW IT!

**STAN'S COOL STUFF**

This bird is named for its behavior of turning over stones to find food. Also called Rock Plover, it can be very tolerant of people when it feeds. The male develops a bare spot on his belly (brood patch) to warm the young, something only females normally have.

# Black-necked Stilt

**Look for the long red-to-pink legs**

**What to look for:**
black and white with ridiculously long red-to-pink legs and a long black bill; upper parts of the head, neck and back are black and lower parts are white

**REAL QUICK**

Size
**14"**

Nest
**GROUND**

Feeder
**NONE**

summer

**Where you'll find them:**
shallow wetlands

**Calls and songs:**
high-pitched "yep-yep" given in rapid series

**On the move:**
slowly stalking through water

**What they eat:**
aquatic insects, small crustaceans, amphibians, snails, and tiny fish

**Nest:**
ground; Mom and Dad build it

**Eggs, chicks, and childcare:**
3–5 off-white eggs with dark markings; Dad incubates the eggs during the day and Mom does at night; both Mom and Dad feed the babies

**Spends the winter:**
in southern states, Mexico, Central America, and South America

**SAW IT!**

## STAN'S COOL STUFF

These are strong and elegant birds that stalk shallow wetlands. They stab at prey or swing their bill from side to side, skimming the water for insects. Populations appear stable, but they are susceptible to their shallow ponds drying up in summer.

American Oystercatcher
Look for the large red-orange bill

**What to look for:**
chunky shorebird with a large red-orange bill and a red eye-ring, black head, a white chest and belly, dark-brown wings, sides, and back, and pink legs

**Where you'll find them:**
at the beach

**Calls and songs:**
gives a series of loud, high-pitched calls that sound like "wheep"

**On the move:**
stabs its bill between the shells of oysters, clams, and other **mollusks** to prevent them from closing, or hammers shells, shattering them with a few powerful blows

**What they eat:**
mollusks, **crustaceans**, shellfish, and worms

**Nest:**
ground nest; Dad and Mom construct it

**Eggs, chicks, and childcare:**
2–4 olive eggs with brown marks; Dad and Mom take turns incubating and doing the childcare

**Spends the winter:**
along coastal Virginia; West Indies, East, Gulf, and South American coasts

**REAL QUICK**

Size
**18–19"**

Nest
**GROUND**

Feeder
**NONE**

year-round

SAW IT!

**STAN'S COOL STUFF**

This large, handsome shorebird stands out at the beach. It uses its heavy, flattened bill to pry open shellfish and probe the sand for mollusks. The young quickly learn the oyster-opening technique from their parents and are soon feeding themselves.

# Pileated Woodpecker

Look for the bright red crest

**What to look for:**
bright-red crest that looks like a hat; bright-red forehead and mustache, and a black back; female has a black forehead and lacks a red mustache

**Where you'll find them:**
prefers areas with lots of woodland

**Calls and songs:**
drums on hollow branches, chimneys, and such to announce territory; loud, rapid "cuk-cuk-cuk" calls carry over a long distance

**On the move:**
white leading edge of wings flashes brightly during flight

**What they eat:**
insects (especially its favorite, carpenter ants); visits **suet** feeders and feeders with peanuts

**Nest:**
cavity in a dead or live tree trunk

**Eggs, chicks, and childcare:**
3–5 white eggs; Mom and Dad sit on the eggs and **regurgitate** bugs to feed the youngsters

**Spends the winter:**
doesn't **migrate**; stays in Virginia year-round

**REAL QUICK**

Size
**19"**

Nest
**CAVITY**

Feeder
**SUET**

year-round

SAW IT!

**STAN'S COOL STUFF**

This is our largest woodpecker. It's shy, despite its size. It digs oval holes up to a few feet long in tree trunks, looking for bugs to eat. You'll see large wood chips at the base of those trees. The young come out of the nest looking and sounding just like the adults.

# Osprey

**Look for the dark line through the eyes**

**What to look for:**
white chest, belly, and head, with a dark eye line

**Where you'll find them:**
always near water, from rivers to wetlands

**Calls and songs:**
a high-pitched, whistle-like **call**, often given in flight as a warning

**On the move:**
can hover for a few seconds before diving to catch a fish; carries fish in a headfirst position in flight for better aerodynamics

**What they eat:**
fish

**Nest:**
platform made with twigs; on a raised wooden platform, man-made tower, or in a tall dead tree

**Eggs, chicks, and childcare:**
2–4 white eggs with brown marks; parents sit on the eggs and feed the **hatchlings**

**Spends the winter:**
in southern coastal states, Mexico, Central and South America

**REAL QUICK**

Size
21–24"

Nest
PLATFORM

Feeder
NONE

summer
migration

SAW IT!

## STAN'S COOL STUFF

The Osprey is the only species in its family. It is the only **raptor** that plunges feetfirst into the water to catch fish. Bald Eagles (pg. 65) will harass it for its catch. At one time, it was almost extinct. It was reintroduced to many regions, and populations are now stable.

# Bald Eagle

**Look for the white head**

JUVENILE

**What to look for:**
white head and tail, curved yellow bill and yellow feet; juvenile has white speckles and a gray bill

**Where you'll find them:**
often near water; likes open areas with daily food

**Calls and songs:**
weak, high-pitched screams, one after another

**On the move:**
a spectacular aerial mating **display**: one eagle flips upside down and locks talons with another; both fall, tumbling to earth, then break apart and fly off

**What they eat:**
fish, **carrion** (dead rabbits and squirrels), birds (mainly ducks)

**Nest:**
massive platform of sticks, usually in a tree; nests used for many years can weigh up to 1,000 pounds

**Eggs, chicks, and childcare:**
2–3 off-white eggs; Mom and Dad share all duties

**Spends the winter:**
in Virginia; moves around in winter to find open water and fish; some migrate to southeastern states

**REAL QUICK**

Size
**31–37"**

Nest
**PLATFORM**

Feeder
**NONE**

year-round
summer
migration
winter

**SAW IT!**

**STAN'S COOL STUFF**

Bald Eagles nearly became extinct, but they're doing well now. Their wingspan is huge, stretching out to 7½ feet! They return to the same nest and add more sticks each year, enlarging it over time. The heads and tails of juveniles turn white at 4–5 years.

# Tree Swallow

**Look for the white chin and chest**

**What to look for:**
blue-green bird with a white chin, chest, and belly, and long, pointed wings

**Where you'll find them:**
ponds, lakes, rivers, and farm fields

**Calls and songs:**
gives a series of gurgles and chirps; chatters when upset or threatened

**On the move:**
flies back and forth across fields, feeding on bugs; uses rapid wingbeats, and then glides; family units gather in large flocks for migration

**What they eat:**
insects—big and small

**Nest:**
cavity; adds grass and lines it with feathers; uses an old woodpecker hole or a wooden nest box

**Eggs, chicks, and childcare:**
4–6 white eggs; Mom sits on the eggs; Mom and Dad bring bugs to feed the babies

**Spends the winter:**
some stay in coastal Virginia; southern coastal states, Mexico, and Central America

**REAL QUICK**

Size
**5–6"**

Nest
**CAVITY**

Feeder
**NONE**

year-round
summer

SAW IT!

**STAN'S COOL STUFF**

This swallow is a good bird to have around because it eats many bugs. You can attract it with a nest box, but it will compete with Eastern Bluebirds (pg. 71) for the cavity. It finds dropped feathers to line its nest and plays with them on its way back to the nest.

# Barn Swallow

**Look for the deeply forked tail**

**What to look for:**
sleek blue-black back, rusty chin, cinnamon belly, and a long, deeply forked tail

**Where you'll find them:**
wetlands, farms, suburban yards, and parks

**Calls and songs:**
gives a twittering **warble** that's followed by a rapid mechanical sound

**On the move:**
flaps continuously, often low over land or water; unlike other swallows, it rarely glides

**What they eat:**
bugs—especially beetles, wasps (caught carefully), and flies

**Nest:**
cup made of mud; brings in up to 1,000 beak-loads of mud to build nest, often on a building; usually it's in a **colony** of 4–6 birds; sometimes nests alone

**Eggs, chicks, and childcare:**
4–5 white eggs with brown marks; Mom sits on the eggs, and Mom and Dad feed the chicks

**Spends the winter:**
in South America

**REAL QUICK**

Size
7"

Nest
CUP

Feeder
NONE

summer

SAW IT!

## STAN'S COOL STUFF

The Barn Swallow is the only swallow in Virginia with a deeply forked tail. It drinks while flying low over water, and it sips the waterdrops on wet leaves. It bathes when it flies through rain or sprinklers. Usually, it nests on a barn, house, or under a bridge.

# Eastern Bluebird

**Look for the rusty-red chest**

MALE

FEMALE

**What to look for:**
sky-blue head, back, wings, and tail, with a rusty-red chest and white belly; female is grayer

**Where you'll find them:**
open habitats (prefers farm fields, pastures, and roadsides), forest edges, parks, and yards

**Calls and songs:**
male repeats a distinctive "chur-lee chur chur-lee" song mostly in spring as he displays to the female

**On the move:**
short flights from tree to tree; often perches in trees or on posts, dropping to ground to grab bugs

**What they eat:**
insects, fruit; visits mealworm and **suet** feeders

**Nest:**
cavity; adds a soft lining in an old woodpecker hole or a bluebird nest box

**Eggs, chicks, and childcare:**
4–5 pale-blue eggs; Mom incubates the eggs, and Dad and Mom feed the kids; 2 broods per year

**Spends the winter:**
in Virginia

**REAL QUICK**

Size
7"

Nest
CAVITY

Feeder
MEALWORM

year-round

SAW IT!

**STAN'S COOL STUFF**

The bluebird is a cousin of the American Robin (pg. 165). It was nearly eliminated due to a lack of tree cavities, but it's thriving now thanks to people who have put up bluebird nest boxes. The young of the first **brood** sometimes help care for the second brood.

# Blue Jay

Look for the large crest

**What to look for:**
vivid blue bird with a black **necklace**; a large crest, which the jay raises and lowers at will

**Where you'll find them:**
in the woods and all around your backyard

**Calls and songs:**
loud, noisy, and mimics other birds; screams like a hawk around feeders to scare away other birds

**On the move:**
carries seeds and nuts in a pouch under its tongue during flight

**What they eat:**
insects, fruit, seeds, nuts, bird eggs, and babies in other nests; visits seed feeders, ground feeders with corn, and any feeder with peanuts

**Nest:**
cup of twigs in a tree, near the main trunk

**Eggs, chicks, and childcare:**
4–5 speckled green-to-blue eggs; Mom sits on the eggs; Mom and Dad feed the little ones

**Spends the winter:**
in Virginia; moves around to find an abundant source of food

**REAL QUICK**

Size
**12"**

Nest
**CUP**

Feeder
**HOPPER**

year-round

**SAW IT!**

## STAN'S COOL STUFF

Blue Jays are very intelligent. They store food in hiding places, called caches, to eat later. They can remember where they hid thousands of nuts! Jays will imitate hawks to scare off other birds at feeders before they land to get the food.

# Belted Kingfisher

Look for the large, ragged crest

MALE

FEMALE

**What to look for:**
broad blue band on a white chest, ragged crest; female has a rusty band below her blue band

**Where you'll find them:**
rarely away from water; usually at banks of rivers, lakes, and large streams

**Calls and songs:**
gives a loud **call** that sounds like a machine gun rattling; mates know each other's call

**On the move:**
flashes white wing patches during flight

**What they eat:**
small fish

**Nest:**
cavity in the bank of a river, lake, or cliff; digs a tunnel up to 4 feet long to the nest chamber

**Eggs, chicks, and childcare:**
6–7 white eggs; Mom and Dad sit on the eggs and feed fish to their youngsters

**Spends the winter:**
in Virginia; moves around to find food

**REAL QUICK**

Size
**12–14"**

Nest
**CAVITY**

Feeder
**NONE**

year-round

SAW IT!

## STAN'S COOL STUFF

Belted Kingfishers perch near water and dive in headfirst to catch fish. Parents drop dead fish into the water to teach their young to dive. Kingfishers have short legs with two toes fused together. This helps a lot when they dig (**excavate**) a burrow for nesting.

# Little Blue Heron

**Look for the black-tipped blue-gray bill**

**What to look for:**
dark blue-to-purple heron with a reddish-purple head and neck, several long plumes on the crown, dull green legs and feet, and a blue-gray bill with a black tip during breeding season; non-breeding bill is gray; juvenile is white with yellowish legs and feet, and a gray bill with a black tip

**Where you'll find them:**
look for it feeding in freshwater lakes, rivers, and ponds, and in saltwater marshes and wetlands

**Calls and songs:**
often silent but gives a deep, hoarse **call** if startled

**On the move:**
stalks **prey** slowly

**What they eat:**
fish and aquatic insects

**Nest:**
platform, in a large **colony** near saltwater sites

**Eggs, chicks, and childcare:**
2–6 light-blue eggs; parents share the childcare

**Spends the winter:**
some don't **migrate**, staying in Virginia all year

REAL QUICK

Size
22–26"

Nest
PLATFORM

Feeder
NONE

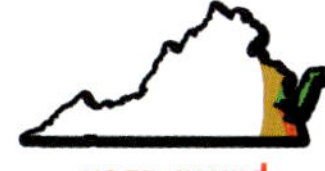

year-round
summer
migration

SAW IT!

**STAN'S COOL STUFF**

This heron is unusual because the young look very different from the adults. They start out with all-white feathers, which makes them look like Snowy Egrets (pg. 207). During the first year, they turn blotchy white. By the second year, they look like the adults.

# Pine Siskin

**Look for the yellow on the wings**

**What to look for:**
brown with a heavily streaked back, chest, and belly, yellow wing bars, and yellow at base of tail

**Where you'll find them:**
**coniferous** to **deciduous** forests, open fields

**Calls and songs:**
gives a series of high-pitched, wheezy calls; also gives a wheezing **twitter**

**On the move:**
moves around to visit feeders in flocks of up to 20 or more birds, often with other finch species; flashes yellow wing markings in flight

**What they eat:**
seeds, bugs; visits seed (especially thistle) feeders

**Nest:**
cup; builds nest in a conifer near the end of a branch, where needles are dense

**Eggs, chicks, and childcare:**
3–4 speckled greenish-blue eggs; Mom incubates the eggs; Mom and Dad bring food to the kiddies

**Spends the winter:**
in Virginia; moves around in search of food during winter

**REAL QUICK**

Size
**5"**

Nest
**CUP**

Feeder
**TUBE OR HOPPER**

winter

**SAW IT!**

## STAN'S COOL STUFF

The Pine Siskin is a finch that breeds in small groups. Nests in the group are often only a few feet apart. The male feeds the female during **incubation**. Juveniles have a yellow tint on their chests and chins, but they lose this by late summer of their first year.

# House Finch

**Look for the heavily streaked chest**

FEMALE

MALE
pg. 197

**What to look for:**
brown bird with heavy streaks on a white chest

**Where you'll find them:**
forests, city, and suburban areas, around homes, parks, and farms

**Calls and songs:**
male sings a loud, cheerful warbling song

**On the move:**
moves around in small family units; never travels long distances

**What they eat:**
seeds, fruit, and leaf buds; comes to seed feeders and feeders with a glop of grape jelly

**Nest:**
cup, but occasionally in a cavity; likes to nest in a hanging flower basket or on a front door wreath

**Eggs, chicks, and childcare:**
4–5 pale-blue eggs, lightly marked; Mom sits on the eggs and Dad feeds her while she incubates; Mom and Dad feed the **brood**

**Spends the winter:**
in Virginia; moves around to find food

**REAL QUICK**

Size
5"

Nest
CUP

Feeder
TUBE OR HOPPER

year-round

SAW IT!

**STAN'S COOL STUFF**

The House Finch is very social and can be the most common bird at feeders. It was introduced to New York from the western US in the 1940s. Now, it's found all across the country. Unfortunately, it suffers from an eye disease that causes the eyes to crust over.

# House Wren

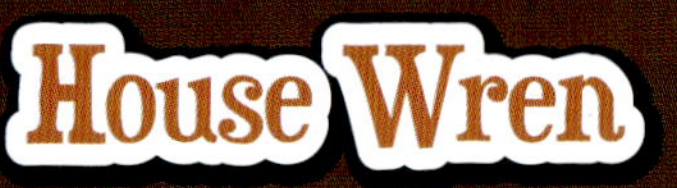

Look for the slightly curved bill

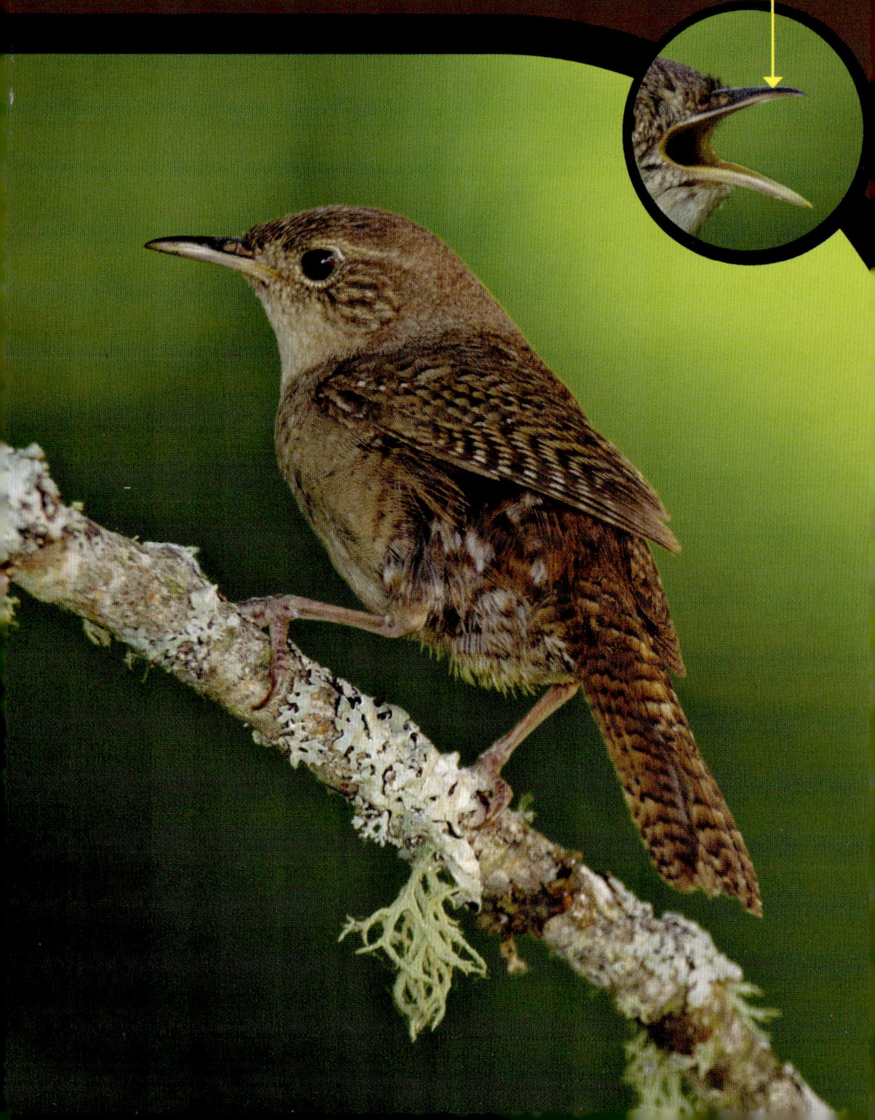

**What to look for:**
brown bird with light-brown marks on the wings and tail, a slightly curved brown bill

**Where you'll find them:**
brushy yards, woodlands, forest edges, and parks

**Calls and songs:**
sings a lot; during the mating season, it sings from dawn to dusk

**On the move:**
short flights from protective bushes; holds its tail up briefly after landing

**What they eat:**
insects, spiders, and snails

**Nest:**
cavity in a tree or birdhouse; easily attracted to a nest box; builds a twiggy nest in spring and lines it with pine needles and grass

**Eggs, chicks, and childcare:**
4–6 tan eggs with brown marks; Mom and Dad incubate the eggs and raise the chicks

**Spends the winter:**
in southern states and Mexico

**REAL QUICK**

Size
5"

Nest
CAVITY

Feeder
NONE

summer

SAW IT!

**STAN'S COOL STUFF**

The male chooses several nest cavities and puts a few small twigs in each. The female selects one cavity and then fills it with short twigs. Often, she will have trouble fitting long twigs through the entrance hole, but she'll try again and again until she's successful.

# Carolina Wren

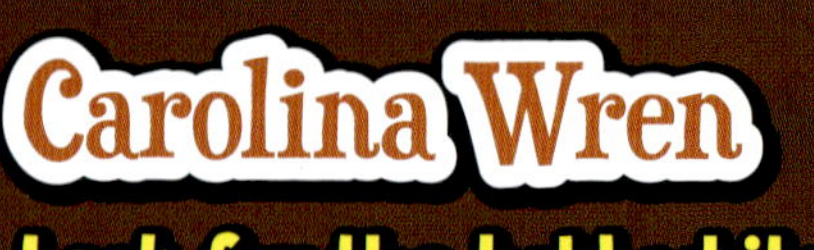

**Look for the bold white eyebrows**

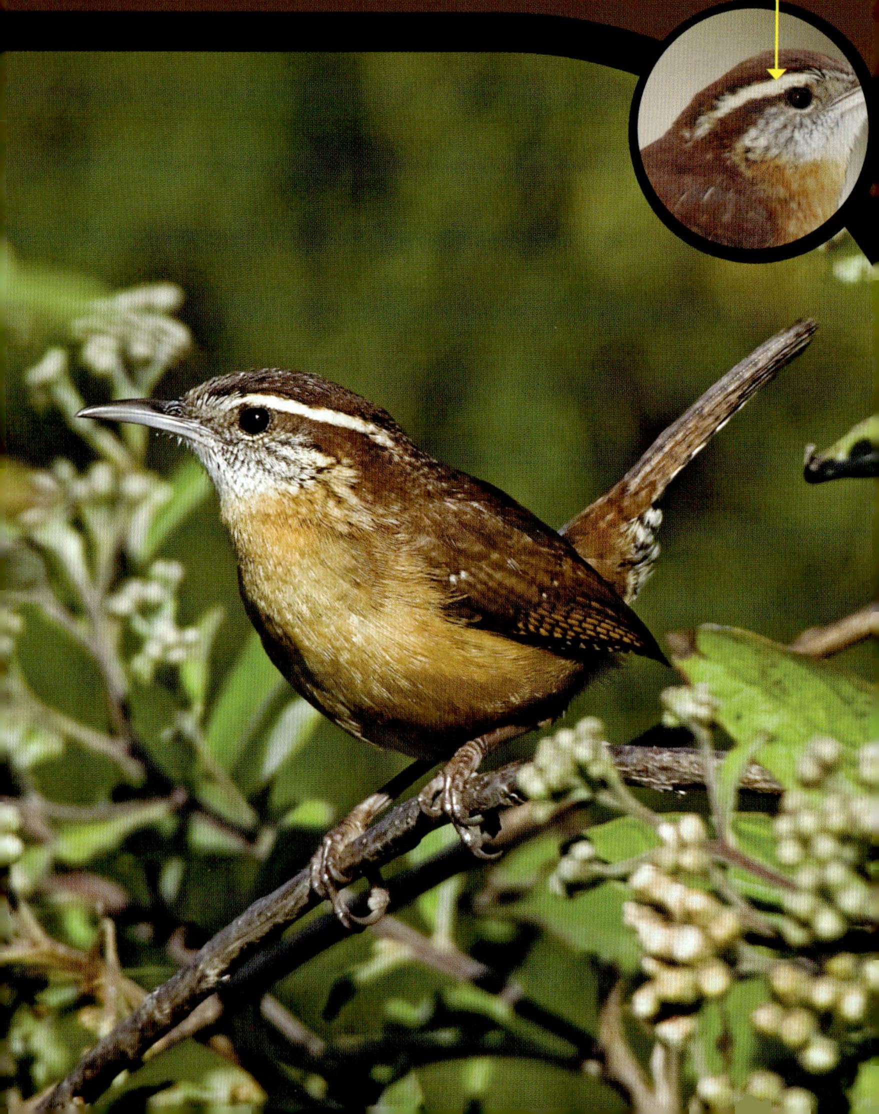

**What to look for:**
orange-yellow chest and belly, a white throat, bold white eyebrows, and a stubby tail, often held up

**Where you'll find them:**
brushy yards and woodlands

**Calls and songs:**
sings year-round; male sings up to 40 song types, singing one song repeatedly before switching to another; female also sings, resulting in duets

**On the move:**
short, fast flights, often perching high up to sing before flying again

**What they eat:**
insects, fruit, and few seeds; comes to **suet** and mealworm feeders

**Nest:**
cavity; nests in birdhouses and in unusual places, like mailboxes, car bumpers, and broken taillights

**Eggs, chicks, and childcare:**
4–6 white (sometimes pink) eggs with brown marks; Mom incubates; parents feed the babies

**Spends the winter:**
doesn't **migrate**; stays in Virginia year-round

**REAL QUICK**

Size
5½"

Nest
CAVITY

Feeder
SUET OR MEALWORM

year-round

SAW IT!

## STAN'S COOL STUFF

Carolina Wrens have a long-term **pair bond.** Mated pairs stay with each other in their territory all year long. They can have up to three broods per year. The male often takes over feeding the young when the female nests again.

# Dark-eyed Junco

**Look for the ivory-to-pink bill**

**What to look for:**
plump bird with a tan-to-brown chest, head, and back, a white belly, and a tiny ivory-to-pink bill

**Where you'll find them:**
on the ground in small flocks with other juncos and sparrows

**Calls and songs:**
a beautiful, loud musical **trill** lasting 2–3 seconds

**On the move:**
outermost tail feathers are white and appear as a white V during flight

**What they eat:**
seeds (scoffs down many weed seeds) and insects; visits ground and seed feeders

**Nest:**
cup on the ground in a wide variety of habitats; female chooses a well-hidden nest site

**Eggs, chicks, and childcare:**
3–5 white eggs with reddish-brown marks; Mom incubates the eggs; Dad and Mom feed the babies

**Spends the winter:**
stays in parts of Virginia; moves around in winter

**REAL QUICK**

Size
5½"

Nest
CUP

Feeder
GROUND

year-round
winter

SAW IT!

**STAN'S COOL STUFF**

The junco is one of our most common winter birds. Females **migrate** farther south than males. This round, dark-eyed bird uses both feet at the same time to "**double-scratch**" the ground, exposing seeds and insects to eat.

# Song Sparrow

**Look for the central dark spot on the chest**

**What to look for:**
brown bird with heavy dark streaking on the chest, some connecting to form a central dark spot

**Where you'll find them:**
scrubby areas, backyards, open fields, forest edges

**Calls and songs:**
sings constantly, repeating a loud, clear song every couple of minutes; starts a song with 3 separate notes, then finishes up with a **trill**; sings from thick shrubs and defends its territory with songs

**On the move:**
rarely in **flocks** (unlike many sparrow species)

**What they eat:**
insects and seeds; rarely visits ground feeders with seeds

**Nest:**
cup; Mom is the sole designer and builder

**Eggs, chicks, and childcare:**
3–4 pale-blue-to-green eggs with reddish-brown splotches; Mom sits on the eggs, and Mom and Dad give food to the **hatchlings**; 2 **broods** per year

**Spends the winter:**
in Virginia; moves around in search of food

**REAL QUICK**

Size
5–6"

Nest
CUP

Feeder
GROUND

year-round

SAW IT!

**STAN'S COOL STUFF**

This bird has many varieties, but they all have a central dark spot on their chests, and they sing unique songs. Song Sparrows are ground feeders. They make little scraping hops backward with both feet to **double-scratch** and uncover seeds to eat.

# House Sparrow

**Look for the black throat patch**

MALE

FEMALE

**What to look for:**
brown back, gray belly and crown, large black patch from throat to chest; female is light brown with distinct light eyebrows, lacks a throat patch

**Where you'll find them:**
just about any **habitat**, from cities to farms

**Calls and songs:**
one of the first birds heard in cities during spring

**On the move:**
nearly always in small flocks

**What they eat:**
seeds, insects, and fruit; comes to seed feeders

**Nest:**
cavity; uses dried grass, scraps of plastic, paper and whatever else is available to construct an oversized domed cup within the cavity

**Eggs, chicks, and childcare:**
4–6 white eggs with brown marks; Mom sits on the eggs; Mom and Dad feed the little ones

**Spends the winter:**
in Virginia; moves around to find food

**REAL QUICK**

Size
6"

Nest
CAVITY

Feeder
TUBE OR HOPPER

year-round

SAW IT!

**STAN'S COOL STUFF**

The House Sparrow is very comfortable being around people. It was introduced to Central Park in New York City from Europe in 1850. It adjusted to nearly all habitats and now is seen across North America. Populations are decreasing in the US and worldwide.

# White-throated Sparrow

**Look for the light stripes on the head**

WHITE-STRIPED

TAN-STRIPED

**What to look for:**
striped head, white or tan throat patch, and small yellow **lores** between the eyes

**Where you'll find them:**
bogs, evergreen, and leafy forests, under feeders

**Calls and songs:**
sings a wonderful song all year and can even be heard at night, sounding like "oh-Canada, Canada"

**On the move:**
often hangs around on the ground with other sparrows during winter

**What they eat:**
insects, seeds, and fruit; comes to ground feeders

**Nest:**
cup on the ground under a small tree

**Eggs, chicks, and childcare:**
4–6 greenish, bluish, or creamy-white eggs with reddish-brown marks; Mom incubates the eggs; Mom and Dad both take care of the babies

**Spends the winter:**
in Virginia, southern states, and Mexico

**REAL QUICK**

Size
6-7"

Nest
CUP

Feeder
GROUND

winter

SAW IT!

**STAN'S COOL STUFF**

This bird has two color variations: white-striped and tan-striped. Both variations mate with each other. Both variations also sing, except for the tan-striped females. This is odd, and scientists aren't sure why those females don't sing. Maybe you can figure it out.

# Eastern Towhee

Look for the rusty sides

pg. 19

REAL QUICK

Size
7–8"

Nest
CUP

Feeder
GROUND

**What to look for:**
light brown bird with rusty sides, a white belly, red eyes, and a long brown tail with a white tip

**Where you'll find them:**
shrubby areas with short trees and thick bushes, backyards and parks

**Calls and songs:**
calls "tow-hee" distinctly; also has a characteristic **call** that sounds like "drink-your-tea"

**On the move:**
short flights between shrubby areas and heavy **cover**; flashes white wing patches during flight

**What they eat:**
insects, seeds, and fruit; comes to ground feeders

**Nest:**
cup; Mom constructs the nest

**Eggs, chicks, and childcare:**
3–4 creamy-white eggs with brown marks; Mom incubates the eggs; Dad and Mom feed the young

**Spends the winter:**
some stay in Virginia; southern states and South America

SAW IT!

**STAN'S COOL STUFF**

The towhee is named for its distinctive "tow-hee" call. It hops backward with both feet, raking leaves to find insects and seeds. In southern coastal states, some have white eyes and others have red eyes. Usually heard well before it is seen.

# Rose-breasted Grosbeak

**Look for the large white eyebrows**

**FEMALE**

**MALE**
pg. 47

**What to look for:**
heavily streaked bird with orange-to-yellow wing linings, bold white eyebrows, and a large ivory bill

**Where you'll find them:**
in the woods, usually in pairs around seed feeders

**Calls and songs:**
female and male sing a rich, robin-like song with a distinctive chip **note** sprinkled in the tune; male is much louder and clearer than the female and sings to her from high up in trees

**On the move:**
short flights from tree to tree with rapid wingbeats; male flashes white wing patches in flight

**What they eat:**
insects, seeds, and fruit; visits seed feeders

**Nest:**
cup in a mature **deciduous** forest, in a small tree

**Eggs, chicks & childcare:**
3–5 speckled blue-green eggs; Mom and Dad sit on the eggs and feed the young chicks; juveniles come with the adults to bird feeders

**Spends the winter:**
in Mexico, Central and South America

**REAL QUICK**

Size
**7–8"**

Nest
**CUP**

Feeder
**TUBE OR HOPPER**

summer
migration

SAW IT!

## STAN'S COOL STUFF

The common name "Grosbeak" refers to the large bill, which the bird uses to crush seeds. In spring, females return to Virginia several days after the males arrive. After the young leave the nest (**fledge**), they come to seed feeders with the adults.

# Brown-headed Cowbird

**Look for the pointed gray bill**

FEMALE

MALE
pg. 21

**What to look for:**
brown bird with a sharp, pointed gray bill

**Where you'll find them:**
forest edges, open fields, farmlands, and backyards

**Calls and songs:**
sings a low, gurgling song that sounds like water moving; cowbird young are raised by other bird parents, but they still end up singing and calling like their own parents, whom they've never heard

**On the move:**
Mom flies quietly to another bird's nest, swiftly lays an egg, then flies quickly away

**What they eat:**
insects and seeds; visits seed feeders

**Nest:**
doesn't nest; lays eggs in the nests of other birds

**Eggs, chicks, and childcare:**
5–7 white eggs with brown marks; the **host** bird incubates any number of cowbird eggs in her nest and feeds the cowbird young along with her own

**Spends the winter:**
in Virginia

**REAL QUICK**

Size
7½"

Nest
NONE

Feeder
TUBE OR HOPPER

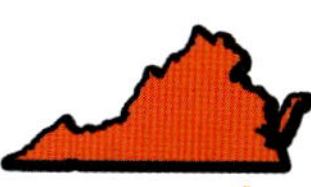

year-round

SAW IT!

**STAN'S COOL STUFF**

Cowbirds are **brood parasites**, meaning they don't nest or raise their own families. Instead, they lay their eggs in other birds' nests, leaving the host birds to raise their young. Cowbirds have laid their eggs in the nests of more than 200 other bird species.

# Cedar Waxwing

**Look for the waxy-looking red wing tips**

**What to look for:**
sleek bird with a pointed crest, black mask, light-yellow belly, and waxy-looking red wing tips; tail has a bold yellow tip; juvenile lacks red wing tips

**Where you'll find them:**
treetops, forest edges, in trees with fruit

**Calls and songs:**
constantly makes a high-pitched "sreee" whistling sound while it's perched or in flight

**On the move:**
flies in flocks of 5–100 birds; moves from area to area, looking for berries

**What they eat:**
berry-like cedar cones, fruit, seeds, and insects

**Nest:**
cup; Mom and Dad construct it together

**Eggs, chicks, and childcare:**
4–6 pale-blue eggs with brown marks; Mom sits on the eggs; Mom and Dad feed the little ones

**Spends the winter:**
in Virginia; wanders around in search of available food supplies

**REAL QUICK**

Size
7½"

Nest
CUP

Feeder
NONE

year-round

SAW IT!

**STAN'S COOL STUFF**

The waxwing is named for its waxy-looking red wing tips and for the cedar's small, blueberry-like cones that it likes to eat. Before berries are abundant, it eats bugs. The young obtain the mask after their first year of life and red wing tips after their second year.

Spotted Sandpiper
Look for white line over the eyes
BREEDING
WINTER

**What to look for:**
olive-brown back with black spots on a white chest and belly; white line over eyes; long, dull-yellow legs; long bill; winter **plumage** lacks spots on the chest and belly

**Where you'll find them:**
along shorelines of large ponds, lakes, and rivers

**Calls and songs:**
gives a rapid series of "weet-weet-weet" calls when frightened and flying away

**On the move:**
flies with arched or cupped wings; rapid wingbeats

**What they eat:**
aquatic insects

**Nest:**
ground; Dad builds it

**Eggs, chicks, and childcare:**
3–4 brownish eggs with brown markings; only Dad sits on the eggs

**Spends the winter:**
in southern states, Mexico, and Central and South America

**REAL QUICK**

Size
8"

Nest
GROUND

Feeder
NONE

summer

SAW IT!

**STAN'S COOL STUFF**

This is one of the most widespread sandpipers in America. When feeding, they are constantly bobbing and teetering. Newly hatched chicks also teeter. It is unknown what function the teetering plays in the bird's life.

Look for the short black bill

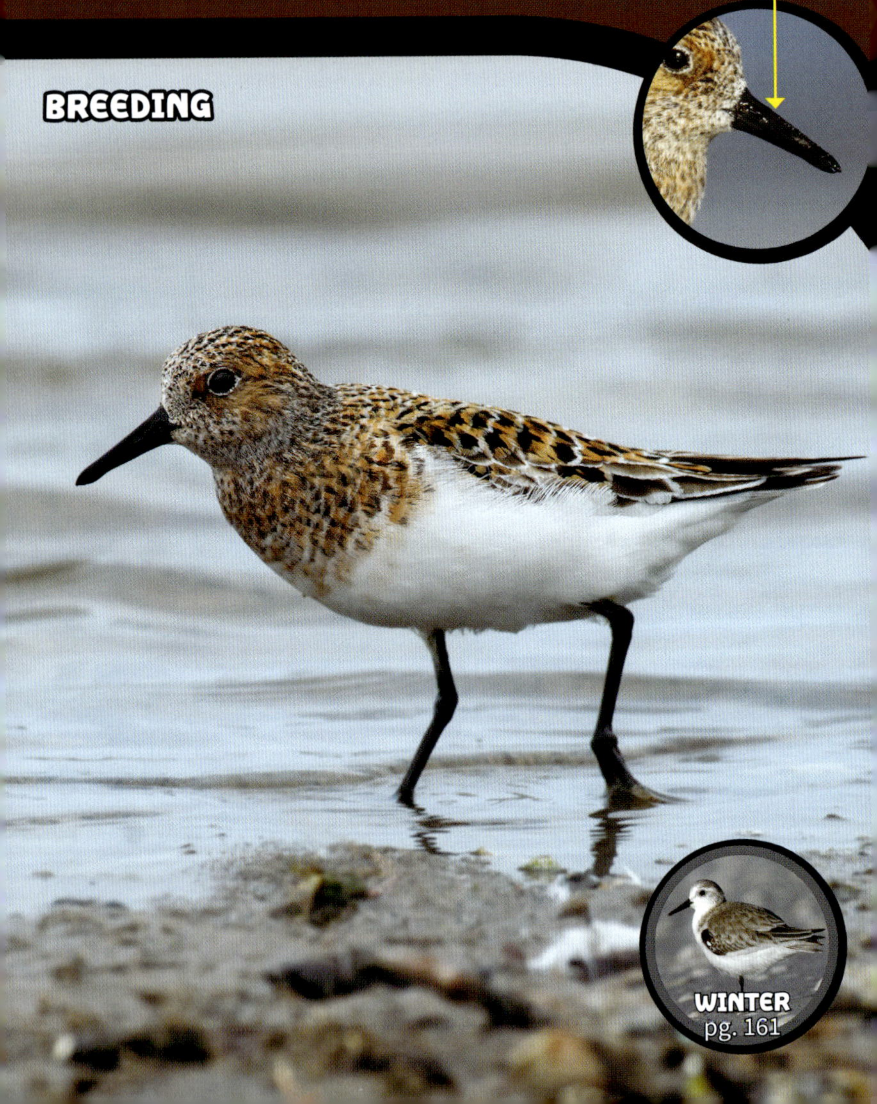

BREEDING

WINTER
pg. 161

**What to look for:**
rusty head, chest, and back with a white belly and black legs and bill; white on the wings, seen only in flight

**Where you'll find them:**
on sandy beaches

**Calls and songs:**
male gives a high-pitched **call** in flight when he displays to the female

**On the move:**
when waves retreat at the beach, groups run out to feed; often hops away from people on one leg; performs a distraction **display** when threatened

**What they eat:**
insects, crabs, worms, and small **mollusks**

**Nest:**
ground nest; Dad builds it

**Eggs, chicks, and childcare:**
3–4 greenish-olive eggs with brown marks; the parents do the **incubation** and feed the kids

**Spends the winter:**
coastal Virginia, Mexico, Central America, and South America

**REAL QUICK**

Size
8"

Nest
GROUND

Feeder
NONE

SAW IT!

## STAN'S COOL STUFF

The Sanderling is one of the most common shorebirds in Virginia. It has breeding **plumage** from April to August and winter plumage from August to April. To rest, it stands on one leg and tucks the other leg into its belly feathers. It nests in the Arctic.

# Northern Cardinal

**Look for the reddish bill**

**What to look for:**
tan-to-brown bird with a black mask and a large reddish bill; juvenile has a blackish-gray bill

**Where you'll find them:**
wide variety of habitats, including backyards and parks; usually likes thick **vegetation**

**Calls and songs:**
calls "whata-cheer-cheer-cheer" in spring; both female and male sing and give chip notes all year

**On the move:**
short flights from **cover** to cover, often landing on the ground

**What they eat:**
loves sunflower seeds and enjoys insects, fruit, peanuts, and **suet**; visits seed feeders

**Nest:**
cup of twigs and bark strips, often low in a tree

**Eggs, chicks, and childcare:**
3–4 speckled bluish-white eggs; Mom and Dad share the incubating and feeding duties

**Spends the winter:**
doesn't **migrate**; gathers with other cardinals and moves around to find good sources of food

**REAL QUICK**

Size
8–9"

Nest
CUP

Feeder
TUBE OR HOPPER

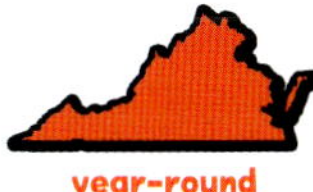

year-round

SAW IT!

**STAN'S COOL STUFF**

Cardinals are sunbathers! Sometimes, they stretch out in the sun, spreading their wings and fanning their tails. They are the first to arrive at feeders in the morning and the last to leave before dark. Females sing as loud as the males—only a few bird species do this.

# Red-winged Blackbird

**Look for the white eyebrows**

pg. 25

**What to look for:**
heavily streaked with a pointed brown bill and white (sometimes yellow) eyebrows

**Where you'll find them:**
around marshes, wetlands, lakes, and rivers

**Calls and songs:**
male sings and repeats calls from cattail tops and the surrounding **vegetation**

**On the move:**
flocks with as many as 10,000 birds gather in autumn, often with other blackbirds

**What they eat:**
seeds in spring and autumn, insects in summer; visits seed and **suet** feeders

**Nest:**
cup in a thick stand of cattails over shallow water

**Eggs, chicks, and childcare:**
3–4 speckled bluish-green eggs; Mom does all the incubating, but both parents feed the babies

**Spends the winter:**
in Virginia; moves around in winter to find food

**REAL QUICK**

Size
$8\frac{1}{2}$"

Nest
CUP

Feeder
TUBE OR HOPPER

year-round

SAW IT!

**STAN'S COOL STUFF**

During autumn and winter, thousands of these birds gather in farm fields, wetlands, and marshes. Come spring, males sing to defend territories and show off their wing patches (**epaulets**) to the females. Later, males can be aggressive when defending their nests.

# American Kestrel

Look for the black lines on the face

MALE

FEMALE

**What to look for:**
rusty back, blue-gray wings, spotted chest, two black lines on the face, a wide black band on the tip of tail; female has rusty wings, dark tail bands

**Where you'll find them:**
open fields, prairies, farm fields, along highways

**Calls and songs:**
loud series of high-pitched "klee-klee-klee" calls

**On the move:**
hovers in midair near roads, then dives for **prey**; pumps tail up and down after landing on a perch

**What they eat:**
bugs (especially grasshoppers), small animals and birds, reptiles

**Nest:**
cavity in a tree or wooden nest box; doesn't add nesting material

**Eggs, chicks, and childcare:**
4–5 white eggs with brown marks; parents take turns sitting on the eggs and feeding the babies

**Spends the winter:**
in Virginia

**REAL QUICK**

Size
9–11"

Nest
CAVITY

Feeder
NONE

year-round

SAW IT!

## STAN'S COOL STUFF

The kestrel is a small falcon that perches nearly upright. The male and female have different markings—this is unusual for a **raptor**. It can see **ultraviolet light**. That ability helps it find mice and other prey by their urine, which glows bright yellow in ultraviolet light.

# Northern Bobwhite

**Look for the white eye stripe and chin**

MALE

FEMALE

**What to look for:**
short, stocky, and mostly brown with short, gray tail, prominent white eye stripe and chin, and reddish-brown sides and belly, often with black lines and dots; female similar to male but with buff-brown eye stripe and chin

**Where you'll find them:**
open fields, shrubby areas

**Calls and songs:**
both male and female give a loud clear **call** that sounds like "bob-white"

**On the move:**
usually runs, but flies short distances close to the ground

**What they eat:**
enjoys insects, seeds and fruit

**Nest:**
ground; depression lined with grass; often pulls nearby **vegetation** over nest to help conceal it

**Eggs, chicks & childcare:**
12–15 white-to-creamy eggs; Mom and Dad sit on the eggs; Mom and Dad feed the babies

**Spends the winter:**
in Virginia; moves around to find food

**REAL QUICK**

Size
10"

Nest
GROUND

Feeder
GROUND

year-round

SAW IT!

**STAN'S COOL STUFF**

Bobwhite population has decreased more than 80 percent in the last 50 years. They prefer to walk or run but are strong fliers. A group of bobwhites is called a "covey" or "bevy." The oldest bob-white recorded was over 6 years old, but most only live 2 or 3 years.

# Killdeer

Look for the two black neckbands

**What to look for:**
brown back, white belly, two black bands around the neck like a **necklace**; a bold reddish-orange rump, visible in flight

**Where you'll find them:**
open country, vacant fields, along railroad tracks, driveways, gravel pits, and wetland edges

**Calls and songs:**
gives a very loud and distinctive "kill-deer" **call**

**On the move:**
fakes a broken wing to draw intruders away from the nest, and then takes flight once the nest is safe

**What they eat:**
loves bugs; also eats worms and snails

**Nest:**
ground; Dad makes just a slight depression in gravel, often very hard to see

**Eggs, chicks, and childcare:**
3–5 tan eggs with brown marks; Dad and Mom incubate the eggs and lead the **hatchlings** to food

**Spends the winter:**
in Virginia

**REAL QUICK**

Size
**11"**

Nest
**GROUND**

Feeder
**NONE**

year-round

**SAW IT!**

## STAN'S COOL STUFF

Scientists group the Killdeer in the family of shorebirds, but you're more likely to spot one along railroad tracks, around farms, and in other dry habitats than you would at the lakeshore. It's the only shorebird with two black neckbands. It migrates in small flocks.

# Northern Flicker

**Look for the black mustache**

MALE

FEMALE

**What to look for:**
brown-and-black bird with a black mustache and black **necklace**, a speckled chest, and a red spot on the nape of neck; female lacks a black mustache

**Where you'll find them:**
forests, small woods, backyards, parks

**Calls and songs:**
gives a loud "wacka-wacka" **call**

**On the move:**
flies in a deep, exaggerated up-and-down pattern, flashing yellow under its wings and tail

**What they eat:**
insects (especially ants and beetles); known to eat the eggs of other birds; visits **suet** feeders

**Nest:**
cavity in a tree or in a nest box that is stuffed with sawdust; often reuses an old nest several times

**Eggs, chicks, and childcare:**
5–8 white eggs; Mom and Dad incubate the eggs and feed the baby woodpeckers

**Spends the winter:**
in Virginia; moves around to find food

**REAL QUICK**

Size
**12"**

Nest
**CAVITY**

Feeder
**SUET**

year-round

**SAW IT!**

## STAN'S COOL STUFF

The Northern Flicker is the only woodpecker to regularly feed on the ground. The male often picks the nest site, usually a natural cavity in a tree. Both parents will pitch in to dig as needed, taking as many as 12 days to finish excavating a hole.

# Mourning Dove

**Look for the shimmering colors on the neck**

**What to look for:**

brown-to-gray bird with shiny, **iridescent** pink and greenish blue on the neck, a gray patch on the head, and black spots on the wings and tail

**Where you'll find them:**

around your seed and ground feeders, open fields

**Calls and songs:**

known for its soft, sad (mournful) cooing

**On the move:**

wind rushes through its wing feathers during takeoff and flight, creating a whistling sound

**What they eat:**

seeds; visits ground and seed feeders

**Nest:**

flimsy platform in a tree, made with twigs; often falls apart in a storm or during high winds

**Eggs, chicks, and childcare:**

2 white eggs; parents incubate the eggs and feed a regurgitated liquid to their young for the first few days of life

**Spends the winter:**

in Virginia; moves around to find food

**REAL QUICK**

Size
**12"**

Nest
**PLATFORM**

Feeder
**GROUND**

year-round

**SAW IT!**

## STAN'S COOL STUFF

This dove is a ground feeder that bobs its head as it walks. It's one of the few birds that drink without lifting its head, like the Rock Pigeon (pg. 171). Parents **regurgitate** a liquid, called **crop-milk**, to feed to their young (**squab**) during their first few days of life.

Boat-tailed Grackle
Look for the golden-brown head
FEMALE
MALE
pg. 33

**What to look for:**
golden-brown head and chest, darker wings, a long dark tail, and bright-yellow eyes

**Where you'll find them:**
coastal saltwater marshes and inland marshes

**Calls and songs:**
noisy, giving several harsh, high-pitched calls and several squeaks

**On the move:**
travels in large flocks with other blackbirds; flight is typically level, not in an up-and-down pattern

**What they eat:**
insects, berries, seeds, grains and fish; comes to seed and **suet** feeders

**Nest:**
cup; Mom makes it with mud or cow dung and grass; nests twice each year in a small **colony**

**Eggs, chicks, and childcare:**
2–4 pale-greenish-blue eggs with brown marks; Mom incubates the eggs and feeds the babies

**Spends the winter:**
doesn't **migrate**; stays in coastal Virginia year-round and moves around to find food

**REAL QUICK**

Size
13–15"

Nest
CUP

Feeder
HOPPER

year-round

SAW IT!

**STAN'S COOL STUFF**

Boat-tails are sometimes seen picking bugs off the backs of cattle. Nests in cattail stands. Forms large flocks in winter. Got its name from the way the male holds its tail in flight, forming a V like the keel of a boat.

# Willet

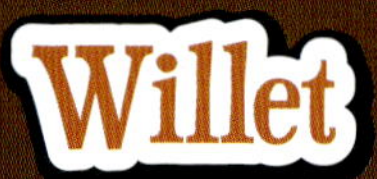

**Look for the boldly patterned wings**

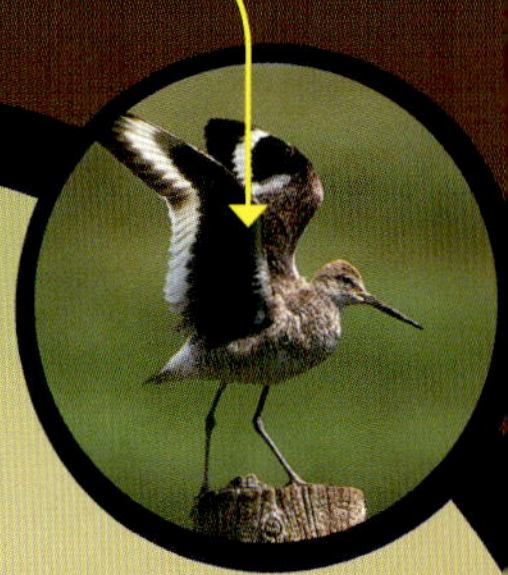

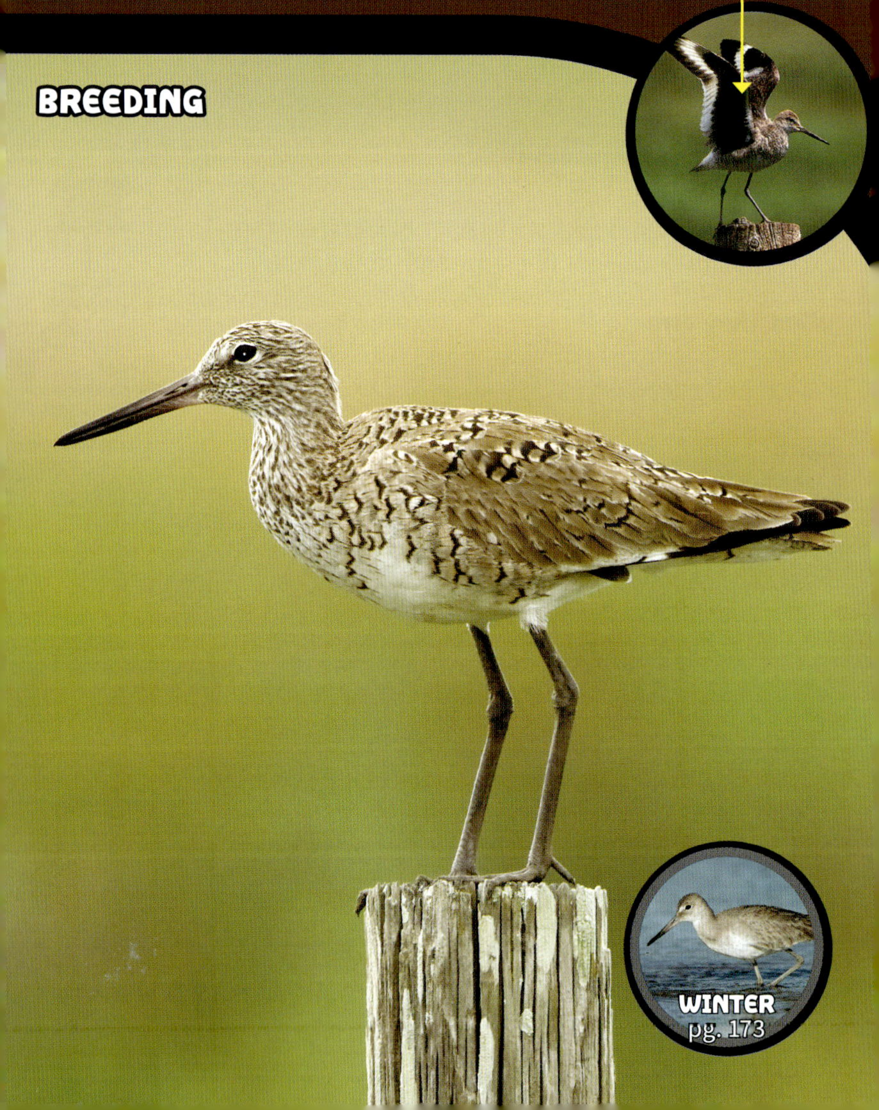
BREEDING
WINTER
pg. 173

**What to look for:**
brown bird with a white belly, and brown legs and bill; distinctive black-and-white pattern on the wings, seen flashing in flight or during **display**

**Where you'll find them:**
at the beach

**Calls and songs:**
calls "pill-will-willet" during the breeding season; gives a "kip-kip-kip" alarm **call** as it takes flight

**On the move:**
easy to identify due to the black-and-white wing pattern that flashes when the bird flaps rapidly

**What they eat:**
insects, small fish, small crabs, worms, and clams

**Nest:**
ground nest; Mom builds the nest

**Eggs, chicks, and childcare:**
3–5 olive eggs with dark marks; parents sit on the eggs and feed the young

**Spends the winter:**
in coastal Virginia

**REAL QUICK**

Size
14–16"

Nest
GROUND

Feeder
NONE

year-round

SAW IT!

**STAN'S COOL STUFF**

This bird is seen along the coast and is very common on beaches all winter. It is a medium-sized sandpiper that uses its long bill to probe into sand in search of food. It nests on the ground along the East and Gulf coasts, in some western states, and in Canada.

# Blue-winged Teal

**Look for the white crescent on the face**

MALE

FEMALE

**What to look for:**

male is brown with black speckles, has a gray head with a white crescent on the face, a white patch on the tail, and a blue wing patch (**speculum**); female is duller and lacks a white facial crescent and white patch on the tail

**Where you'll find them:**

wetlands and lakes

**Calls and songs:**

male makes a high-pitched squeak; female quacks

**On the move:**

direct flight to and from water; **flocks** fly fast in tight formation; female performs a **display** to draw predators away from the nest and young

**What they eat:**

aquatic plants, seeds, and aquatic bugs

**Nest:**

cozy ground nest some distance from the water

**Eggs, chicks, and childcare:**

8–11 creamy-white eggs; Mom does the **incubation** and feeds the ducklings

**Spends the winter:**

in coastal Virginia and southern states, Mexico, and Central America

**REAL QUICK**

Size
**15–16"**

Nest
**GROUND**

Feeder
**NONE**

SAW IT!

**STAN'S COOL STUFF**

This is one of the smallest ducks in North America. It **migrates** farther than most other ducks, nesting as far north as Alaska. The name "Blue-winged" refers to its blue wing patch, which is easiest to see when the bird is in flight.

# Red-shouldered Hawk

**Look for the reddish shoulders**

**What to look for:**
cinnamon-red head, shoulders, chest, and belly; brown wings and back with white spots, and a long tail with black-and-white bands; reddish wing linings, seen in flight

**Where you'll find them:**
wooded backyards, forest edges, woodlands

**Calls and songs:**
extremely vocal; gives distinctive, loud screams

**On the move:**
alternates flapping with gliding

**What they eat:**
reptiles, amphibians, large insects, and small birds

**Nest:**
large platform made of sticks and lined with sprigs of evergreen or other soft materials; usually in a fork of a large tree

**Eggs, chicks, and childcare:**
2–4 white eggs with dark marks; Mom and Dad sit on the eggs and provide for the youngsters

**Spends the winter:**
in Virginia; moves around to find food

REAL QUICK

Size
16–19"

Nest
CAVITY

Feeder
NONE

year-round

SAW IT!

STAN'S COOL STUFF

A common hawk in Virginia. It likes to hunt along forest edges and will search for snakes, frogs, bugs, and other **prey** as it perches. It stays in the same territory for many years. The parents start to build a nest in March. The young leave the nest (**fledge**) by July.

# Wood Duck

Look for the bright white eye-ring

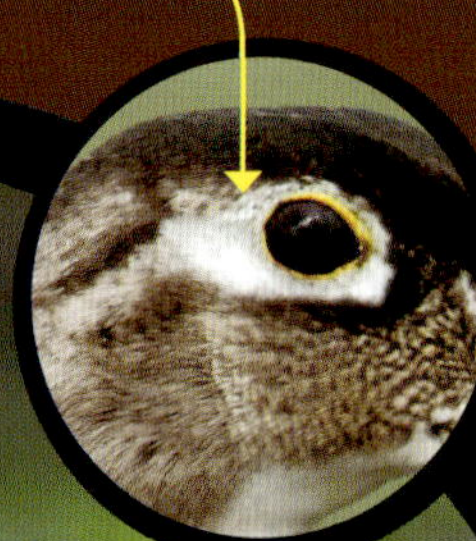

FEMALE

MALE
pg. 189

**What to look for:**
brown with a bold white eye-ring; crest on head and blue wing patch (**speculum**) are harder to see

**Where you'll find them:**
quiet, shallow ponds, and deep in the woods, high up on tree branches

**Calls and songs:**
female calls "oo-eek, oo-eek" when startled and at takeoff; male calls a zipper-like "zeeeet"

**On the move:**
blasts off from the water with loud calls and noisy wings; flies quickly through forest; enters cavity nest from full flight; small, tight group flights

**What they eat:**
aquatic insects, plants, and seeds

**Nest:**
cavity; adds a lining of soft, downy feathers in an old woodpecker hole or a nest box

**Eggs, chicks, and childcare:**
10–15 creamy-white eggs; only Mom incubates the eggs and shows the kids how to feed

**Spends the winter:**
some stay in Virginia

**REAL QUICK**

Size
17–20"

Nest
CAVITY

Feeder
NONE

year-round

SAW IT!

**STAN'S COOL STUFF**

This is a small dabbling duck. The female will lay some eggs in a neighbor's nest (**egg dumping**), sometimes resulting in 20 or more eggs in a nest! The young stay in the nest for a day, then jump from as high as 60 feet to the ground or water and follow their mom.

# Northern Shoveler

Look for the large, shovel-like bill

FEMALE

MALE
pg. 191

**What to look for:**
brown duck with black speckles, green wing mark (**speculum**), and a super-large, spoon-shaped bill

**Where you'll find them:**
shallow wetlands, ponds, and small lakes

**Calls and songs:**
female gives a classic quack; male gives a crazy-sounding combination of popping and quacking, calling "puk-puk, puk-puk, puk-puk"

**On the move:**
swims in tight circles, stirring up insects to eat; small flocks of 5–10 birds swim with bills pointing toward the water; flocks fly in tight formation

**What they eat:**
enjoys aquatic insects; likes plants, too

**Nest:**
ground; Mom gathers plant material and forms it into a circle a short distance from the water

**Eggs, chicks, and childcare:**
9–12 olive eggs; Mom sits on the eggs and leads her little shovelers to food

**Spends the winter:**
in Virginia, southern states, Mexico, and Central America

**REAL QUICK**

Size
19–21"

Nest
GROUND

Feeder
NONE

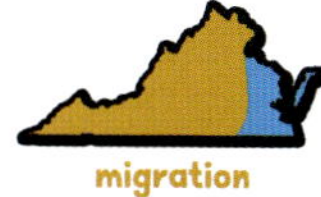

migration
winter

SAW IT!

**STAN'S COOL STUFF**

The Northern Shoveler is a medium-sized duck. It is the only shoveler species found in North America. The name "Shoveler" refers to its peculiar, shovel-like bill. It feeds by using its bill to sift tiny aquatic insects and plants floating on the water's surface.

# Mallard

Look for the orange-and-black bill

FEMALE

MOTTLED DUCK

MALE
pg. 193

**What to look for:**
overall brown duck with an orange-and-black bill, a white tail, and a blue-and-white wing mark (**speculum**), seen best in flight

**Where you'll find them:**
lakes and ponds, rivers and streams, and maybe even your backyard

**Calls and songs:**
the sound a duck makes is based on the female Mallard's classic quack; the male doesn't quack

**On the move:**
sometimes in huge flocks with hundreds of ducks; mostly in small flocks of 6–10, especially in spring

**What they eat:**
seeds, aquatic plants, and insects; visits ground feeders offering corn

**Nest:**
ground; Mom builds it from plants nearby

**Eggs, chicks, and childcare:**
7–10 greenish-to-whitish eggs; Mom incubates the eggs and leads the young to food

**Spends the winter:**
in Virginia; moves around to find food

**REAL QUICK**

Size
**19–21"**

Nest
**GROUND**

Feeder
**GROUND**

year-round

**SAW IT!**

**STAN'S COOL STUFF**

This is a dabbling duck, tipping forward in shallow water to feed on aquatic plants on the bottom. Only the female quacks. It will return to its birthplace each year. The Mottled Duck (see inset) looks very much like the female Mallard except for the bill color.

# Red-tailed Hawk

**Look for the rusty-red tail**

**What to look for:**
**plumage** varies but it's often brown with a white chest, brown belly band, and a rusty-red tail

**Where you'll find them:**
just about anywhere; open country, where it flies over open fields and roadsides; cities, where it perches on freeway light posts, fences, and trees

**Calls and songs:**
gives a high-pitched scream that trails off

**On the move:**
hunts in flight, flying in circles as it searches for **prey**; migrates during the day

**What they eat:**
mice and other animals, birds, snakes, large bugs

**Nest:**
large platform made of sticks, lined with materials such as evergreen needles; often in a large tree

**Eggs, chicks, and childcare:**
2–3 white eggs, sometimes speckled; parents sit on the eggs and provide for the youngsters

**Spends the winter:**
in Virginia; moves around in winter to find food

**REAL QUICK**

Size
19–23"

Nest
PLATFORM

Feeder
NONE

year-round

SAW IT!

**STAN'S COOL STUFF**

This **raptor** is a large hawk with a wide variety of colors from bird to bird, ranging from chocolate to nearly all white. The red color of the tail develops in the second year of life and usually is best seen from above. It returns to the same nest site each year.

# Barred Owl

**Look for the dark eyes**

**What to look for:**
brown-to-gray owl with dark-brown eyes, dark horizontal bars on upper chest, vertical streaks on the lower chest and belly; yellow bill and feet

**Where you'll find them:**
dense woodlands

**Calls and songs:**
gives calls of 6–8 hoots, sounding something like "who-who-who-cooks-for-you"

**On the move:**
a smooth and silent flight, gliding on flat, outstretched wings; often hunts during the day, perching and watching for mice and other **prey**

**What they eat:**
small mammals, birds, fish, reptiles, amphibians

**Nest:**
natural cavity in a tree or uses a nest box with a large entrance hole; doesn't add nesting material

**Eggs, chicks, and childcare:**
2–3 white eggs; Mom sits on the eggs; Mom and Dad attend to the babies

**Spends the winter:**
doesn't **migrate**; stays in Virginia year-round

**REAL QUICK**

Size
**20–24"**

Nest
**CAVITY**

Feeder
**NONE**

year-round

SAW IT!

**STAN'S COOL STUFF**

The Barred Owl is the only common dark-eyed owl in Virginia. It's a chunky bird with a large head. It fishes by hovering over water and then reaches down to grab one. After the young **fledge**, they stay with their parents for up to four months.

# Great Horned Owl

**Look for the feather tufts on the head**

**What to look for:**
"eared" owl with large yellow eyes, a V-shaped white throat, and horizontal barring on the chest

**Where you'll find them:**
just about any **habitat** throughout Virginia

**Calls and songs:**
calls a familiar "hoo-hoo-hoo-hoooo"

**On the move:**
flies silently on big wings that stretch out to 4 feet; takes a few quick flaps, and then glides

**What they eat:**
small to medium mammals, birds (especially ducks), snakes, and insects

**Nest:**
no nest; takes over the nest of another bird or uses a broken tree stump or other semi-cavity

**Eggs, chicks, and childcare:**
2–3 white eggs, laid in January and February; Mom incubates; Dad and Mom feed the **hatchlings**

**Spends the winter:**
doesn't **migrate** and usually hangs around the same area year after year

**REAL QUICK**

Size
**21–25"**

Nest
**NONE**

Feeder
**NONE**

year-round

**SAW IT!**

**STAN'S COOL STUFF**

The "**horns**" of the Great Horned are feather tufts, not ears. Its eyelids close from the top down, like ours. It has fabulous hearing and can hear a mouse moving under a deep pile of leaves. It's one of the few animals that will kill a skunk or a porcupine.

# Wild Turkey

Look for the bare blue-and-red head

MALE

FEMALE

**What to look for:**
funny-looking brown-and-bronze bird, bare blue-and-red head, long thin beard, large fanning tail; female is thinner, duller, and often lacks a beard

**Where you'll find them:**
just about any **habitat**, from suburban yards to prairies and forests

**Calls and songs:**
a fast, descending "gobble-gobble-gobble-gobble" that's often heard before the bird is seen

**On the move:**
a strong flier that can approach 60 miles per hour; also able to fly straight up, and then away

**What they eat:**
insects, seeds, and fruit

**Nest:**
ground; Mom scrapes out a shallow depression and pads it with soft leaves

**Eggs, chicks, and childcare:**
10–12 whitish eggs with dull brown marks; Mom sits on the eggs and leads the babies to food

**Spends the winter:**
moves around Virginia to find **cover** and food

**REAL QUICK**

Size
36–48"

Nest
GROUND

Feeder
NONE

year-round

SAW IT!

**STAN'S COOL STUFF**

The turkey is the largest game bird in Virginia. The male's head and neck change color when displaying for females. It sees three times better than people, and it can hear sounds from a mile away. A male will lead a group of up to 20 females.

# Brown Pelican

**Look for the huge gray bill**

NON-BREEDING

**What to look for:**
gray-brown body with a black belly, a very long gray bill, a white or yellow head, and dark rust on the back of the neck during breeding season; non-breeding back of the neck is white

**Where you'll find them:**
at the beach

**Calls and songs:**
silent; snaps the upper bill and lower bill together to make a loud popping sound

**On the move:**
often sits on posts at beachside docks

**What they eat:**
fish; occasionally amphibians and eggs

**Nest:**
ground nest, in a large **colony**, often on an island

**Eggs, chicks, and childcare:**
2–4 white eggs; Mom and Dad incubate the eggs and share the childcare

**Spends the winter:**
along coastal Virginia

**REAL QUICK**

Size
46–50"

Nest
GROUND

Feeder
NONE

SAW IT!

## STAN'S COOL STUFF

This bird is common in Virginia, although it was endangered not long ago due to the use of pesticides. It hunts by diving headfirst into the ocean. Then it opens its bill and catches fish with the expandable pouch on the bottom of its bill, like a net.

# Brown-headed Nuthatch

**Look for the brown head**

**What to look for:**
tiny bird with a brown head, dull white chin, chest, and belly, and a gray back

**Where you'll find them:**
open pine forests, woodlands, parks, backyards, forest edges

**Calls and songs:**
a characteristic spring **call**, "whi-whi-whi-whi," given during February and March

**On the move:**
creeps down tree trunks and branches headfirst, searching for hidden insects and insect eggs; makes quick, short flights from tree to tree

**What they eat:**
insects and seeds; visits seed and **suet** feeders

**Nest:**
cavity; parents **excavate** or will use a vacant woodpecker hole, a natural cavity or a nest box

**Eggs, chicks, and childcare:**
3–5 white eggs with dark marks; Mom sits on the eggs; Mom and Dad feed the young

**Spends the winter:**
doesn't **migrate**; moves around to find food

**REAL QUICK**

Size
4½"

Nest
CAVITY

Feeder
TUBE OR SUET

year-round

SAW IT!

## STAN'S COOL STUFF

This is one of 17 nuthatch species worldwide. It has an extra-long hind toe claw, called a nail, on each foot, giving it the ability to cling to trees and climb down headfirst. Works hard to remove seeds from pinecones. Pairs defend a very small territory all year.

# Black-capped Chickadee

Look for the black cap

**What to look for:**
mostly gray bird with a black cap and throat patch, tan sides and belly, and a white chest

**Where you'll find them:**
nearly all habitats—just look around for this bird

**Calls and songs:**
calls "chika-dee-dee-dee-dee"; also gives a high-pitched, two-toned "fee-bee" **call** during spring; can have different calls in different regions

**On the move:**
flies short distances with short, fluttery wings

**What they eat:**
seeds, bugs, and fruit; visits seed and **suet** feeders

**Nest:**
excavates a cavity or uses a nest box; gathers mostly green moss for the nest and fur to line it

**Eggs, chicks, and childcare:**
5–7 white eggs with fine brown marks; Mom and Dad sit on the eggs and feed their **brood**

**Spends the winter:**
doesn't **migrate**; moves around to find food and shelter

**REAL QUICK**

Size
5"

Nest
CAVITY

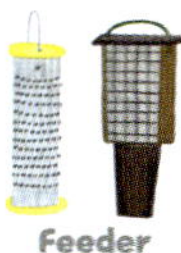

Feeder
TUBE OR SUET

year-round

SAW IT!

**STAN'S COOL STUFF**

You can attract this bird with a seed feeder or nest box. Usually, it's the first to find a new feeder. It is easily tamed and hand-fed. Most of the diet comes from bird feeders, so it can be a common urban bird. It's often seen with nuthatches, woodpeckers, and other birds.

# Carolina Chickadee

**What to look for:**
mostly gray bird with a black cap and throat patch, tan sides and belly, and a whitish chest

**Where you'll find them:**
nearly all habitats—just look around for this bird

**Calls and songs:**
calls "chika-dee-dee-dee-dee"; also gives a high-pitched, two-toned "fee-bee" **call** during spring; can have different calls in different regions

**On the move:**
flies short distances with short, fluttery wings

**What they eat:**
seeds, bugs, and fruit; visits seed and **suet** feeders

**Nest:**
excavates a cavity or uses a nest box; gathers mostly green moss for the nest and fur to line it

**Eggs, chicks, and childcare:**
5–7 white eggs with reddish-brown marks; Mom and Dad sit on the eggs and feed their **brood**

**Spends the winter:**
doesn't **migrate**; moves around to find food and shelter instead

**REAL QUICK**

Size
5"

Nest
CAVITY

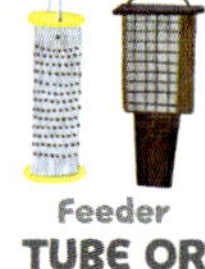

Feeder
TUBE OR SUET

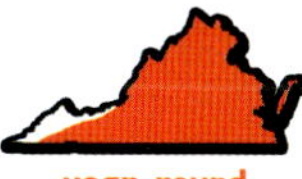

year-round

SAW IT!

**STAN'S COOL STUFF**

You can attract this bird with a seed feeder or nest box. Usually, it's the first to find a new feeder. It's easily tamed and hand-fed. Much of its diet comes from bird feeders, so it can be a common urban bird. It's often seen with nuthatches, woodpeckers, and other birds.

# White-breasted Nuthatch

**Look for the white chest**

MALE

FEMALE

**What to look for:**
gray back with a white face, chest, and belly, a black cap and nape of neck, and a large white patch on the rump; female has a gray cap and **nape**

**Where you'll find them:**
woodlands, parks, backyards, forest edges

**Calls and songs:**
a characteristic spring **call**, "whi-whi-whi-whi," given during February and March

**On the move:**
climbs down tree trunks headfirst, looking for hidden bugs; quick, short flights from tree to tree

**What they eat:**
bugs, bug eggs, seeds; visits seed and **suet** feeders

**Nest:**
cavity; Mom and Dad build a nest in an empty woodpecker hole or a natural cavity

**Eggs, chicks, and childcare:**
5–7 white eggs with brown marks; Mom sits on the eggs; Mom and Dad feed the little ones

**Spends the winter:**
doesn't **migrate** in Virginia; moves around to find food

**REAL QUICK**

Size
5–6"

Nest
CAVITY

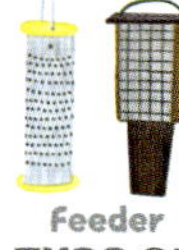

Feeder
TUBE OR SUET

year-round

SAW IT!

## STAN'S COOL STUFF

This is one of almost 30 nuthatch species worldwide. It has an extra-long hind toe claw, called a nail, on each foot, giving it the ability to cling to trees and climb down headfirst. It's often in flocks with chickadees. Pairs stay together all year and defend their territory.

# Yellow-rumped Warbler

**Look for the bright yellow patches**

MALE

FEMALE

FIRST WINTER

**What to look for:**
gray with black streaks on the chest and yellow patches on head, flanks, and rump; female is duller gray; first-winter juvenile is similar to the female

**Where you'll find them:**
can be seen in any **habitat** during migration; seems to prefer **deciduous** forests and forest edges

**Calls and songs:**
sings a wonderful song in spring; calls a single robust "chip," heard mostly during migration

**On the move:**
quickly moves among trees and from the ground to trees; flits around upper branches of tall trees

**What they eat:**
insects and berries; visits **suet** feeders in spring

**Nest:**
cup; Mom builds the nest on her own in forests

**Eggs, chicks, and childcare:**
4–5 white eggs with brown marks; Mom sits on the eggs; Mom and Dad feed the young

**Spends the winter:**
in Virginia

REAL QUICK

Size
5–6"

Nest
CUP

Feeder
SUET

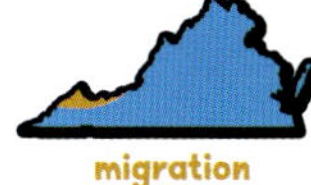

migration
winter

SAW IT!

## STAN'S COOL STUFF

This bird is also called the Myrtle Warbler. One of the most common warblers in the state. The male molts his gray feathers in fall, changing to a dull color like the female for the winter. He keeps his yellow patches all year.

# Dark-eyed Junco

**Look for the pink bill**

MALE

FEMALE
pg. 87

**What to look for:**
plump bird with a gray-to-charcoal chest, head, and back, a white belly, and a tiny pink bill

**Where you'll find them:**
on the ground in small flocks with other juncos and sparrows

**Calls and songs:**
a beautiful, loud musical **trill** lasting 2–3 seconds

**On the move:**
outermost tail feathers are white and appear as a white V during flight

**What they eat:**
seeds (scoffs down many weed seeds) and insects; visits ground and seed feeders

**Nest:**
cup on the ground in a wide variety of habitats; female chooses a well-hidden nest site

**Eggs, chicks, and childcare:**
3–5 white eggs with reddish-brown marks; Mom incubates the eggs; Dad and Mom feed the babies

**Spends the winter:**
stays in parts of Virginia; moves around in winter

**REAL QUICK**

Size
5½"

Nest
CUP

Feeder
GROUND

year-round
winter

SAW IT!

## STAN'S COOL STUFF

The junco is one of our most common winter birds. Females **migrate** farther south than males. This round, dark-eyed bird uses both feet at the same time to "**double-scratch**" the ground, exposing seeds and insects to eat.

# Tufted Titmouse

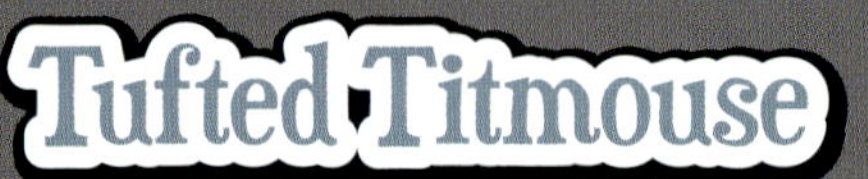

**Look for the pointed crest**

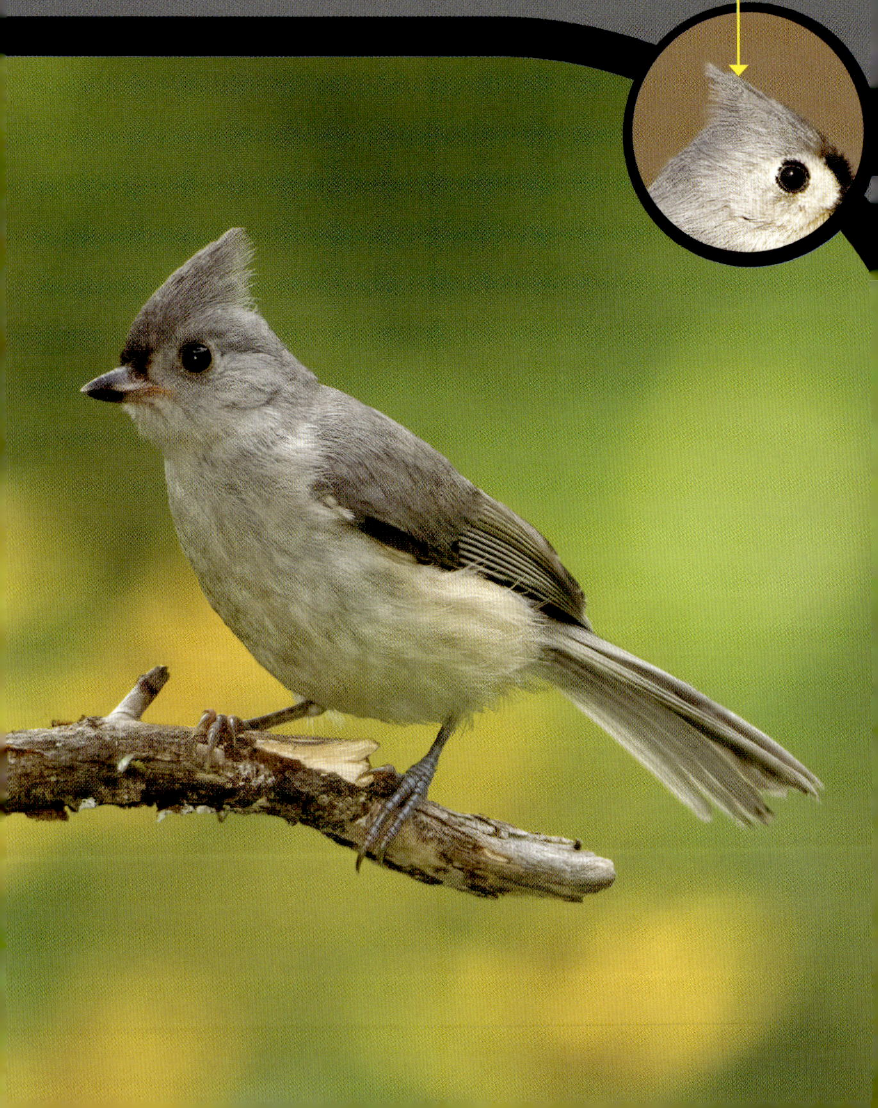

**What to look for:**
gray bird with a white chest and belly, a rusty-brown wash on the flanks, and a pointed crest

**Where you'll find them:**
woodlands, backyards, and parks

**Calls and songs:**
quickly repeats a loud "peter-peter-peter" **call**

**On the move:**
usually seen only one or two at a time, never in big flocks, moving through thick forests and along forest edges

**What they eat:**
insects, seeds (especially black oil sunflower seeds), and fruit; visits seed and **suet** feeders

**Nest:**
cavity; boldly pulls hair from sleeping dogs, cats, or squirrels and uses it to line an old woodpecker hole or a nest box

**Eggs, chicks, and childcare:**
5–7 white eggs with brown marks; Mom sits on the eggs, and Mom and Dad feed the young

**Spends the winter:**
doesn't **migrate**; stays in Virginia year-round

**REAL QUICK**

Size
6"

Nest
CAVITY

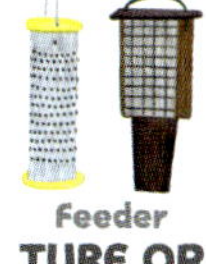

Feeder
TUBE OR SUET

year-round

SAW IT!

## STAN'S COOL STUFF

The titmouse is a common feeder bird. You can attract it by filling a feeder with black oil sunflower seeds or peanuts or by putting up a nest box. The male feeds the female during courtship and nesting. Its name means "small bird."

# Eastern Phoebe

**Look for the tail pumping while perching**

**What to look for:**
plain gray with darker wings, a light olive-green belly, and a thin dark bill

**Where you'll find them:**
end of dead branches in forests, yards, and farms

**Calls and songs:**
repeats a characteristic "fee-bee" **call** over and over from the top of dead branches

**On the move:**
waits on a branch for an insect to pass by, flies out to snatch it, and returns to the branch—called **hawking**; bobs its tail up and down when perched

**What they eat:**
insects

**Nest:**
cup made of mud, grass, and green moss, lined with hair (and sometimes feathers); under a bridge or eave of a house, or in another sheltered area

**Eggs, chicks, and childcare:**
4–5 white eggs; Mom sits on the eggs to incubate; Dad and Mom feed their **hatchlings**

**Spends the winter:**
in coastal Virginia; to southern states and Mexico

**REAL QUICK**

Size
7"

Nest
CUP

Feeder
NONE

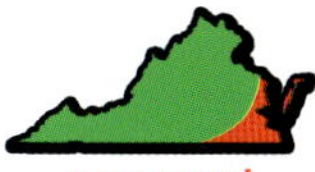

year-round
summer

SAW IT!

## STAN'S COOL STUFF

The Eastern Phoebe is named for its very distinct call, "fee-bee." Because it is a plain bird that doesn't have distinctive markings, it's easier to identify by its hawking and tail-pumping behaviors, and by the call that sounds like its name.

**Look for the gray back**

**WINTER**

**What to look for:**
a light-colored sandpiper with a gray head and back, white belly, and black legs and bill; white stripe on the wings, seen only in flight

**Where you'll find them:**
on sandy beaches

**Calls and songs:**
gives a fast, high-pitched squeak just before taking flight and just after takeoff

**On the move:**
when waves retreat at the beach, groups run out to feed; often hops away from people on one leg; performs a distraction **display** when threatened

**What they eat:**
insects, crabs, worms, and small **mollusks**

**Nest:**
ground nest; Dad builds it

**Eggs, chicks, and childcare:**
3–4 greenish-olive eggs with brown marks; the parents do the **incubation** and feed the kids

**Spends the winter:**
in coastal Virginia, Mexico, Central America, and South America

**REAL QUICK**

Size
8"

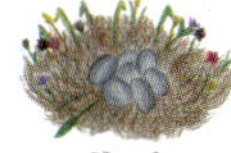

Nest
GROUND

Feeder
NONE

migration
winter

SAW IT!

**STAN'S COOL STUFF**

The Sanderling is one of the most common shorebirds in Virginia. It has winter **plumage** from August to April and breeding plumage from April to August. To rest, it stands on one leg and tucks the other leg into its belly feathers. It nests in the Arctic.

# Gray Catbird

**What to look for:**
gray bird with a black crown and long, thin black bill; often lifts its tail, exposing a chestnut patch

**Where you'll find them:**
thick shrubs, forest edges, backyards, parks

**Calls and songs:**
gives a rasping **call** that sounds like a house cat meowing; often mimics other birds

**On the move:**
quickly zips back into shrubs if approached

**What they eat:**
insects and occasional fruit; visits suet feeders

**Nest:**
cup of small twigs in thick shrubs; often adds a scrap of plastic to the twigs

**Eggs, chicks, and childcare:**
4–6 blue-green eggs; Mom incubates the eggs; Mom and Dad take turns feeding the chicks, but because the parents look the same, it's hard to tell who's doing the feeding!

**Spends the winter:**
to southern states; doesn't **migrate** in eastern Virginia

**REAL QUICK**

Size
9"

Nest
CUP

Feeder
SUET

year-round
summer

SAW IT!

## STAN'S COOL STUFF

This handsome, secretive bird is more often heard than seen. The Chippewa Indian name for it means "the bird that cries with grief." Once you've heard the call, you won't forget it. If a cowbird lays an egg in a catbird nest, the catbird will break it and eject it.

# American Robin

**Look for the rusty-red breast**

MALE

FEMALE

**What to look for:**
black head and a rich, rusty-red breast; female is duller with a gray head and lighter breast

**Where you'll find them:**
loves to hop on lawns in search of worms

**Calls and songs:**
chips and chirps; sings all night in spring; studies report that city robins sing louder than country robins so they can be heard over traffic and noise

**On the move:**
found all over the US in an amazing range of habitats from sea level to mountaintops

**What they eat:**
insects, fruit, and berries, as well as earthworms

**Nest:**
cup; weaves plant materials and uses mud to plaster the nest to a sheltered location

**Eggs, chicks, and childcare:**
4–7 pale-blue eggs; Mom sits on the eggs; Mom and Dad feed the baby robins

**Spends the winter:**
doesn't **migrate** in Virginia

**REAL QUICK**

Size
**9–11"**

Nest
**CUP**

Feeder
**NONE**

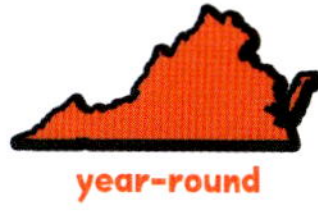

year-round

**SAW IT!**

**STAN'S COOL STUFF**

When a robin walks across your lawn and turns its head to the side, it isn't listening for worms—it is looking for them. Because its eyes are on the sides of its head, a robin must focus its sight out of one eye to see the dirt moving caused by a moving worm.

# Northern Mockingbird

**Look for the white wing patches**

**What to look for:**
silver-gray head and back, a light-gray chest and belly, white wing patches, a mostly black tail with white outer tail feathers

**Where you'll find them:**
on top of a shrub, where it sits for long periods; parks and yards

**Calls and songs:**
imitates or mocks other birds (vocal mimicry); young males often sing at night

**On the move:**
very lively, spreading its wings, flashing its white wing patches, and wagging its tail; wing patches flash during flight or **display**

**What they eat:**
insects and fruit

**Nest:**
cup; Mom and Dad work together to build it

**Eggs, chicks, and childcare:**
3–5 speckled blue-green eggs; Mom sits on the eggs to incubate; Mom and Dad feed their young

**Spends the winter:**
in Virginia

**REAL QUICK**

Size
**10"**

Nest
**CUP**

Feeder
**NONE**

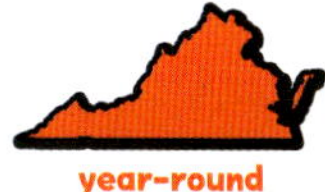

year-round

**SAW IT!**

## STAN'S COOL STUFF

Mockingbirds perform a fantastic mating dance. Pairs hold up their heads and tails and run toward each other. They flash their wing patches and then retreat to nearby **cover**. Usually, they're not afraid of people, so you may be able to get a close look.

# Eurasian Collared-Dove

**Look for the black collar on the back of neck**

**What to look for:**
pale-gray bird with a slightly darker back, wings, and tail, and a black collar on the **nape** of neck; tail is long and squared off at the end

**Where you'll find them:**
wherever it can scratch for seeds

**Calls and songs:**
gives a series of coos, often during a **display**

**On the move:**
usually found in small to large flocks; fast flaps, then glides with wings in a V

**What they eat:**
seeds and fruit; visits ground and seed feeders

**Nest:**
platform on a building, balcony, barn or shed, or under a bridge; Mom and Dad construct it

**Eggs, chicks, and childcare:**
2 creamy-white eggs; Mom and Dad sit on the eggs and **regurgitate** a liquid food, called **crop-milk**, to feed to the young (**squab**) during their first few days of life

**Spends the winter:**
doesn't **migrate**; stays in Virginia year-round

**REAL QUICK**

Size
12½"

Nest
PLATFORM

Feeder
GROUND

year-round

SAW IT!

## STAN'S COOL STUFF

This non-native dove is originally from Asia. People brought it into the Bahamas, and then it flew into Florida in the 1980s. Since then, it has spread across the country. Scientists believe that this bird will continue to expand its range across more of North America.

# Rock Pigeon

Look for the gleaming, iridescent patches

**What to look for:**
color pattern varies; usually shades of gray with gleaming, **iridescent** patches of green mixed with blue; often has a light rump patch

**Where you'll find them:**
nearly anyplace where it can scratch for seeds

**Calls and songs:**
a series of coos, usually given during a **display**

**On the move:**
typically in small to large flocks; flaps rapidly, then glides with wings in a V shape

**What they eat:**
seeds and fruit; visits ground and seed feeders

**Nest:**
platform on a building, balcony, barn or shed, or under a bridge

**Eggs, chicks, and childcare:**
1–2 white eggs; Mom and Dad sit on the eggs and **regurgitate** a liquid, called **crop-milk**, to feed to the young (**squab**) for their first few days of life

**Spends the winter:**
doesn't **migrate**; stays in Virginia year-round

REAL QUICK

Size
13"

Nest
PLATFORM

Feeder
GROUND

year-round

SAW IT!

## STAN'S COOL STUFF

The pigeon was introduced to North America by the early settlers from Europe. Years of breeding in captivity have made it one of the few birds that has a variety of colors. It's also one of the few birds that can drink without tilting its head back.

Look for the boldly patterned wings

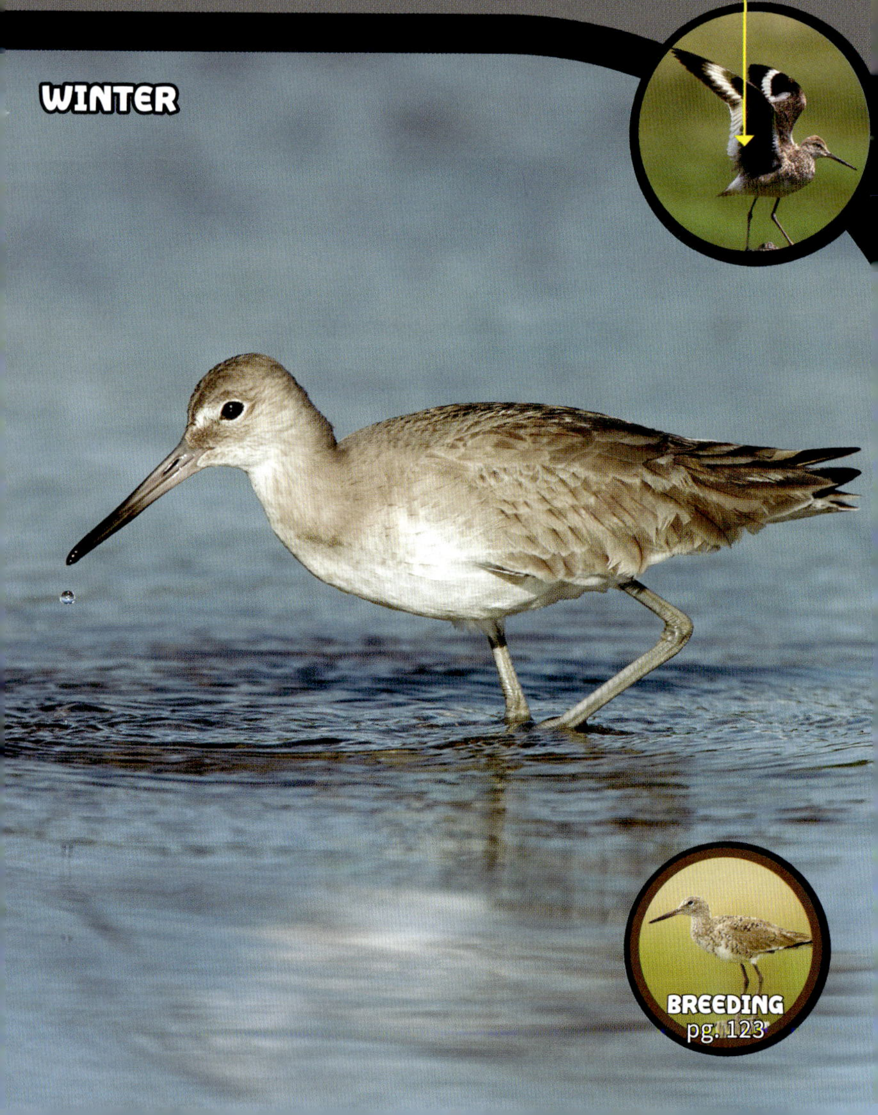

WINTER

BREEDING
pg. 123

**What to look for:**
gray bird with a white belly, and gray legs and bill; distinctive black-and-white pattern on the wings, seen flashing in flight or during **display**

**Where you'll find them:**
at the beach

**Calls and songs:**
calls "pill-will-willet" during the breeding season; gives a "kip-kip-kip" alarm **call** as it takes flight

**On the move:**
easy to identify due to the black-and-white wing pattern that flashes when the bird flaps rapidly

**What they eat:**
insects, small fish, small crabs, worms, and clams

**Nest:**
ground nest; Mom builds the nest

**Eggs, chicks, and childcare:**
3–5 olive eggs with dark marks; parents sit on the eggs and feed the young

**Spends the winter:**
in coastal Virginia

**REAL QUICK**

Size
**14–16"**

Nest
**GROUND**

Feeder
**NONE**

year-round

**SAW IT!**

## STAN'S COOL STUFF

This bird is seen along the coast and is very common on beaches all winter. It is a medium-sized sandpiper that uses its long bill to probe into sand in search of food. It nests on the ground along the East and Gulf coasts, in some western states, and in Canada.

# Sharp-shinned Hawk

**Look for the banded, squared tail**

**What to look for:**
small hawk with a gray back and head, a rusty-red chest, short, rounded wings, red eyes, and a long tail with several dark bands and a squared tip

**Where you'll find them:**
just about any **habitat**, from backyards and parks to woodlands

**Calls and songs:**
gives a loud, high-pitched "kik-kik-kik-kik" **call**

**On the move:**
swoops into backyards toward birds at feeders, often chasing the birds as they flee

**What they eat:**
birds and small mammals

**Nest:**
platform made with sticks, usually high in a tree; Mom constructs it

**Eggs, chicks, and childcare:**
4–5 white eggs with brown marks; Mom sits on the eggs; Mom and Dad feed the little ones

**Spends the winter:**
in Virginia

**REAL QUICK**

Size
**10–14"**

Nest
**PLATFORM**

Feeder
**NONE**

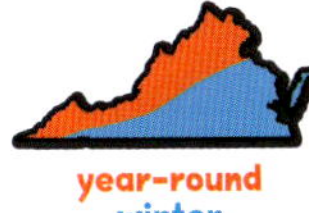

year-round
winter

**SAW IT!**

## STAN'S COOL STUFF

A short wingspan and long tail help this small hawk zoom around trees in pursuit of **prey.** "Sharp-shinned" refers to the sharp edge on the front of its shin. A bird's shin is below its ankle, not above the ankle, like ours. Most birds have a rounded shin.

# Cooper's Hawk

**Look for the banded, rounded tail**

**What to look for:**
gray back, rusty chest, short wings, dark-red eyes, and a long tail with black bands and a rounded tip

**Where you'll find them:**
variety of habitats, from woodlands to backyards and parks

**Calls and songs:**
calls a loud, clear "cack-cack-cack-cack"

**On the move:**
flies with a few quick flaps followed by long glides

**What they eat:**
small birds (hunts birds at feeders) and mammals

**Nest:**
platform made with sticks and leaves, secured high in a tree; Dad and Mom build it together

**Eggs, chicks, and childcare:**
2–4 greenish eggs with brown marks; parents take turns sitting on the eggs and feeding the young

**Spends the winter:**
stays year-round in Virginia; moves around to find food

**REAL QUICK**

Size
**14–20"**

Nest
**PLATFORM**

Feeder
**NONE**

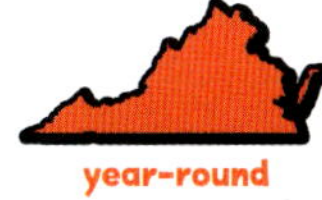

year-round

**SAW IT!**

## STAN'S COOL STUFF

The stubby wings help this medium-sized hawk move around trees while it chases smaller birds. It will ambush **prey**, flying into brush and running after the birds that flee. **Fledglings** have gray eyes that turn yellow at 1 year and dark red after 3–5 years.

# Peregrine Falcon

**Look for the yellow eye-ring**

**What to look for:**
dark gray back, tan-to-white chest, dark **hood** head marking, wide black mustache; yellow eye-ring and base of bill and legs; horizontal bars on belly, legs, and undertail; female is similar to male but noticeably larger

**Where you'll find them:**
canyons, open country and cities with tall buildings

**Calls and songs:**
alarm **call** is a series of high-pitched screams

**On the move:**
fast direct flight with constant wingbeats

**What they eat:**
birds, such as Rock Pigeons in cities and shorebirds and **waterfowl** in rural areas

**Nest:**
ground (scrape) on a cliff edge, tall building, bridge, or smokestack

**Eggs, chicks, and childcare:**
3–4 white eggs with brown marks; parents take turns sitting on the eggs and feeding the young

**Spends the winter:**
some stay in Virginia; moves around to find food in winter, joined by migrants from further north

**REAL QUICK**

Size
16–20"

Nest
GROUND

Feeder
NONE

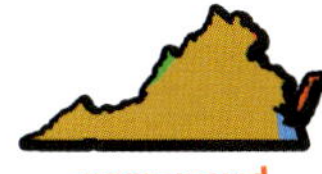

year-round
summer
migration
winter

SAW IT!

**STAN'S COOL STUFF**

Peregrine Falcons were once nearly eliminated due to indiscriminate shooting and trapping. Now, they are doing well in most parts of the country. This is the fastest bird on the planet. It can fly 70 miles per hour in level flight and can dive up to 200 miles per hour.

Canada Goose
Look for the white cheek strap

**What to look for:**
large gray goose with a black neck and head, and a white chin and cheek strap

**Where you'll find them:**
wetlands, ponds, lakes, rivers, and just about any **habitat** with some water

**Calls and songs:**
belts out its classic "honk-honk-honk," especially during flight

**On the move:**
flies in a **flock** in a large V shape when traveling long distances

**What they eat:**
aquatic plants, insects, and seeds

**Nest:**
ground nest of **vegetation** formed into a mound, usually very near or on water

**Eggs, chicks, and childcare:**
5–10 white eggs; Mom incubates the eggs; young follow the parents around and learn what to eat

**Spends the winter:**
in Virginia; moves around to find open water in winter

**REAL QUICK**

Size
25–43"

Nest
GROUND

Feeder
NONE

year-round

SAW IT!

**STAN'S COOL STUFF**

Males guard the flock, bobbing their heads and hissing whenever people approach. Adults **molt** their flight feathers while raising the young, making families temporarily flightless. Canada Geese start to breed in their third year of life. Adults mate for many years.

Great Blue Heron
Look for the long yellow bill

**What to look for:**
tall gray heron with black eyebrows that end in plumes off the back of the head, neck feathers that drop down like a **necklace**, and a long yellow bill

**Where you'll find them:**
open water, from small ponds to large lakes

**Calls and songs:**
when startled, it barks repeatedly like a dog and keeps at it while flying away

**On the move:**
holds its neck in an S shape in flight and slightly cups its wings, trailing its legs straight out behind

**What they eat:**
small fish, frogs, insects, snakes, and baby birds

**Nest:**
platform in a tree near or over open water, in a **colony** of up to 100 birds

**Eggs, chicks, and childcare:**
3–5 blue-green eggs; parents incubate the eggs and feed the **brood**

**Spends the winter:**
doesn't **migrate** in Virginia

**REAL QUICK**

Size
42–48"

Nest
PLATFORM

Feeder
NONE

year-round

SAW IT!

**STAN'S COOL STUFF**

This is the tallest heron in Virginia and very common. It stalks fish in shallow water and strikes at mice, squirrels, and anything else it can capture on land. Red-winged Blackbirds (pg. 25) often attack it to prevent it from taking their babies out of their nests.

# Ruby-throated Hummingbird

**Look for the gleaming ruby-red throat**

MALE

FEMALE

**What to look for:**
tiny **iridescent** green bird with a black throat patch that shimmers bright ruby-red in direct sunlight; female lacks the throat patch

**Where you'll find them:**
many habitats, from yards and parks to forests

**Calls and songs:**
will chatter or buzz to communicate; doesn't sing or hum a melody—it's the incredibly fast flapping wings that create the humming sound

**On the move:**
the only bird that can fly backward; also hovers in midair and flies straight up and straight down!

**What they eat:**
**nectar** and insects; visits nectar feeders

**Nest:**
stretchy cup of plant materials and spiderwebs; glues bits of **lichen** to the outside for camouflage

**Eggs, chicks, and childcare:**
2 white eggs; Mom does all egg and chick care

**Spends the winter:**
in southern states, Mexico, and Central America

**REAL QUICK**

Size
3–3½"

Nest
CUP

Feeder
NECTAR

summer

SAW IT!

**STAN'S COOL STUFF**

This is the tiniest bird in the state, with about the same weight as a US penny. It flaps 50–60 times per second or more when flying. It breathes 250 times per minute, and its heart beats 1,260 times per minute! It feeds at colorful tube-shaped flowers.

# Green Heron

**What to look for:**
short, stocky heron with a blue-green back, a dark-green crest, and a rusty-red chest and neck; short orange legs turn yellow after the breeding season

**Where you'll find them:**
ponds, wetlands, small lakes, and rivers

**Calls and songs:**
often gives an explosive, rasping "skyew" **call** when startled

**On the move:**
waits on the shore or wades stealthily, hunting for food; makes short, quick flights across the water

**What they eat:**
small fish, aquatic insects, amphibians

**Nest:**
platform of sticks in a **coniferous** or **deciduous** tree, often a short distance from water and can be very high in the tree

**Eggs, chicks, and childcare:**
2–4 light-green eggs; parents share the childcare

**Spends the winter:**
in South America; some move to Mexico and Central America

**REAL QUICK**

Size
**16–22"**

Nest
**PLATFORM**

Feeder
**NONE**

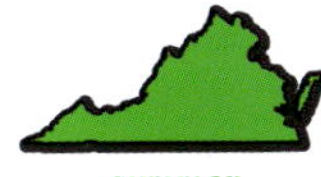

summer

**SAW IT!**

## STAN'S COOL STUFF

The Green Heron holds its head very close to its body. When it's excited, it raises its crest. To attract fish to catch, it will place an object, such as an insect, on the surface of the water. Baby herons make a loud ticking sound, like the ticktock of a clock.

# Wood Duck

Look for the boldly patterned head

pg. 129

**What to look for:**
boldly patterned head and crest with bold white outlines; rusty chest and white belly

**Where you'll find them:**
quiet, shallow ponds and deep in the woods, high up on tree branches

**Calls and songs:**
male calls a zipper-like "zeeeet"; female calls "oo-eek, oo-eek" loudly when startled and at takeoff

**On the move:**
blasts off from the water with loud calls and noisy wings; flies quickly through forest; enters cavity nest from full flight; small, tight group flights

**What they eat:**
aquatic insects, plants, and seeds

**Nest:**
cavity; adds a lining of soft, downy feathers in an old woodpecker hole or a nest box

**Eggs, chicks, and childcare:**
10–15 creamy-white eggs; only Mom incubates the eggs and shows the kids how to feed

**Spends the winter:**
some stay in Virginia

**REAL QUICK**

Size
17-20"

Nest
CAVITY

Feeder
NONE

year-round

SAW IT!

## STAN'S COOL STUFF

This is a small dabbling duck. The female will lay some eggs in a neighbor's nest (**egg dumping**), sometimes resulting in 20 or more eggs in a nest! The young stay in the nest for 24 hours and then jump down to follow their mother. They never return to the nest.

# Northern Shoveler

**Look for the large, shovel-like bill**

MALE

FEMALE

pg. 131

**What to look for:**
shiny, **iridescent** green head with rusty sides, a white chest, and a super-large, spoon-shaped bill

**Where you'll find them:**
shallow wetlands, ponds, and small lakes

**Calls and songs:**
male gives a crazy-sounding combination of popping and quacking, calling "puk-puk, puk-puk, puk-puk"; female gives a classic quack **call**

**On the move:**
swims in tight circles, stirring up insects to eat; small flocks of 5–10 birds swim with bills pointing toward the water; flocks fly in tight formation

**What they eat:**
enjoys aquatic insects; likes plants, too

**Nest:**
ground; Mom gathers plant material and forms it into a circle a short distance from the water

**Eggs, chicks, and childcare:**
9–12 olive eggs; Mom sits on the eggs and leads her little shovelers to food

**Spends the winter:**
in Virginia, southern states, Mexico, and Central America

**REAL QUICK**

Size
19-21"

Nest
GROUND

Feeder
NONE

migration
winter

SAW IT!

## STAN'S COOL STUFF

The Northern Shoveler is a medium-sized duck. It is the only shoveler species found in North America. The name "Shoveler" refers to its peculiar, shovel-like bill. It feeds by using its bill to sift tiny aquatic insects and plants floating on the water's surface.

Mallard
Look for the green head
MALE
FEMALE
pg. 133

**What to look for:**
green head with a white **necklace**, rusty-brown chest, gray sides, yellow bill, orange legs, and feet

**Where you'll find them:**
lakes and ponds, rivers and streams, and maybe even your backyard

**Calls and songs:**
the male doesn't quack; when you think of how a duck sounds, it's based on the female Mallard's classic loud quack

**On the move:**
sometimes in huge flocks with hundreds of ducks; mostly in small flocks of 6–10, especially in spring

**What they eat:**
seeds, aquatic plants, and insects; visits ground feeders offering corn

**Nest:**
ground; Mom builds it from plants nearby

**Eggs, chicks, and childcare:**
7–10 greenish-to-whitish eggs; Mom incubates the eggs and leads the young to food

**Spends the winter:**
in Virginia; moves around to find food

**REAL QUICK**

Size
19–21"

Nest
GROUND

Feeder
GROUND

year-round

SAW IT!

**STAN'S COOL STUFF**

This is a dabbling duck. It tips forward in shallow water to eat plants on the bottom. The male has black feathers in the center of its tail that curl upward. The common name "Mallard" means "male" and refers to the males, which don't help raise their young.

Baltimore Oriole
Look for the black head
MALE
FEMALE
pg. 219

**What to look for:**
flaming orange bird with a black head and back, and black wings with white wing bars

**Where you'll find them:**
parks, yards, and forests; in treetops, where it feeds on caterpillars

**Calls and songs:**
a fantastic songster, singing loudly; often heard before it is seen

**On the move:**
often returns to the same area year after year

**What they eat:**
insects, fruit, and **nectar**; comes to nectar, orange-half, and grape jelly feeders

**Nest:**
pendulous; an interesting nest that looks like a sock hanging from an outer branch of a tall tree

**Eggs, chicks, and childcare:**
4–5 bluish eggs with brown marks; Mom sits on the eggs; Mom and Dad do the childcare

**Spends the winter:**
in Mexico, Central and South America

**REAL QUICK**

Size
7-8"

Nest
PENDULOUS

Feeder
NECTAR

summer
migration

SAW IT!

## STAN'S COOL STUFF

Orioles come to feeders that offer sugar water (nectar), orange halves, or grape jelly. Parents bring their young to feeders. Young males turn orange and black at 1½ years. Members of the blackbird family; closely related to grackles and Red-winged Blackbirds.

# House Finch

**Look for the reddish face and the brown cap**

MALE

YELLOW MALE

FEMALE
pg. 81

**What to look for:**
red-to-orange face, throat, chest, and rump, and a brown cap

**Where you'll find them:**
forests, city, and suburban areas, around homes, parks, and farms

**Calls and songs:**
male sings a loud, cheerful warbling song

**On the move:**
moves around in small family units; never travels long distances

**What they eat:**
seeds, fruit, and leaf buds; comes to seed feeders and feeders with a glop of grape jelly

**Nest:**
cup, but occasionally in a cavity; likes to nest in a hanging flower basket or on a front door wreath

**Eggs, chicks, and childcare:**
4–5 pale-blue eggs, lightly marked; Mom sits on the eggs and Dad feeds her while she incubates; Mom and Dad feed the **brood**

**Spends the winter:**
in Virginia; moves around to find food

**REAL QUICK**

Size
5"

Nest
CUP

Feeder
TUBE OR HOPPER

year-round

**STAN'S COOL STUFF**

The House Finch is very social and is found across the country. It can be the most common bird at feeders. Unfortunately, it suffers from an eye disease that causes the eyes to crust over. It's rare to see a yellow male; yellow **plumage** may be a result of a poor diet.

# Northern Cardinal

**Look for the black mask**

**What to look for:**
all-red bird except for a black mask, and a large red crest and bill

**Where you'll find them:**
wide variety of habitats including backyards and parks; usually likes thick **vegetation**

**Calls and songs:**
calls "whata-cheer-cheer-cheer" in spring; both male and female sing and give chip notes all year

**On the move:**
short flights from **cover** to cover, often landing on the ground

**What they eat:**
loves sunflower seeds and enjoys insects, fruit, peanuts, and **suet**; visits seed feeders

**Nest:**
cup of twigs and bark strips, often low in a tree

**Eggs, chicks, and childcare:**
3–4 speckled bluish-white eggs; Mom and Dad share the incubating and feeding duties

**Spends the winter:**
doesn't **migrate**; gathers with other cardinals and moves around to find good sources of food

**REAL QUICK**

Size
8–9"

Nest
CUP

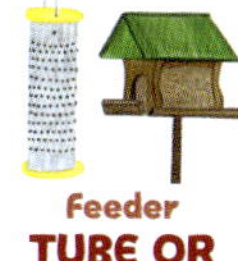

Feeder
TUBE OR HOPPER

year-round

SAW IT!

**STAN'S COOL STUFF**

Cardinals are sunbathers! Sometimes, they stretch out in the sun, spreading their wings and fanning their tails. They're the first to arrive at feeders in the morning and the last to leave before dark. They are territorial and fight their own reflections in windows.

**Look for the black head "hood"**

**BREEDING**

**What to look for:**
black head "**hood**" with a white neck, chest, and belly, gray wings with black wing tips, and orange bill; winter **plumage** head is gray and white with a black bill

**Where you'll find them:**
along the coasts; freshwater and saltwater sites

**Calls and songs:**
a loud series of calls that sound like laughter; male tosses his head back and calls to attract a mate

**On the move:**
almost always in groups, moving from one water **habitat** to another

**What they eat:**
fish, insects on land and in the water

**Nest:**
ground nest, lined with grass, sticks, and rocks; nests in a marsh in a large **colony**

**Eggs, chicks, and childcare:**
2–4 olive eggs with brown marks; parents sit on the eggs and **regurgitate** food to feed the young

**Spends the winter:**
in Virginia

**REAL QUICK**

Size
16–17"

Nest
GROUND

Feeder
NONE

SAW IT!

## STAN'S COOL STUFF

It takes this gull a few years to get adult plumage. During the first year, the young start out mostly brown-and-gray. They look like adults during the second year, but they don't have an all-black head "hood." Juveniles get the breeding plumage in the third year.

# Ring-billed Gull

**Look for the black ring on the bill**

Mostly White

**What to look for:**
white gull with gray wings and a yellow bill with a black ring near the tip; winter **plumage** has speckles on the head and neck

**Where you'll find them:**
shores of large lakes and rivers; often at garbage dumps and parking lots

**Calls and songs:**
calls out a wide variety of loud, rising squawks and squeals—classic gull sounds

**On the move:**
strong flight with constant wing flaps

**What they eat:**
insects and fish; it also picks through garbage, scavenging for other food

**Nest:**
ground; defends a small area around it

**Eggs, chicks, and childcare:**
2–4 off-white eggs with brown marks; Mom and Dad take turns incubating the eggs and feeding their young

**Spends the winter:**
in Virginia, southern coastal states, and Mexico

**REAL QUICK**

Size
18–20"

Nest
GROUND

Feeder
NONE

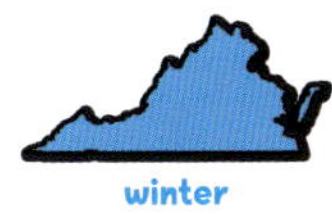

winter

SAW IT!

## STAN'S COOL STUFF

This is one of the most common gulls in the country. Hundreds of these birds often **flock** together. The ring on the bill appears after the first winter. In the fall of the first 3 years, the birds have a different plumage. In the third year, they grow adult plumage.

**Look for the large orange-red bill**

**What to look for:**
large white bird with heavy orange bill and prominent black cap

**Where you'll find them:**
along the shore

**Calls and songs:**
a loud, raspy rattle given repeatedly

**On the move:**
strong flight with constant wing flaps

**What they eat:**
fish and aquatic insects

**Nest:**
ground; mom and dad build it

**Eggs, chicks, and childcare:**
1–2 white eggs with brown marks; Mom and Dad take turns incubating the eggs and feeding their young

**Spends the winter:**
on the coasts of Florida, Mexico, and Central and South America

**REAL QUICK**

Size
20"

Nest
GROUND

Feeder
NONE

SAW IT!

**STAN'S COOL STUFF**

Prefers sandy beaches with shallow coastal waters. It dives head-first into water to catch small fish. Nests on sandy beaches. Young leave nest within 24 hours of hatching and gather together, known as a creche (nursery).

# Snowy Egret

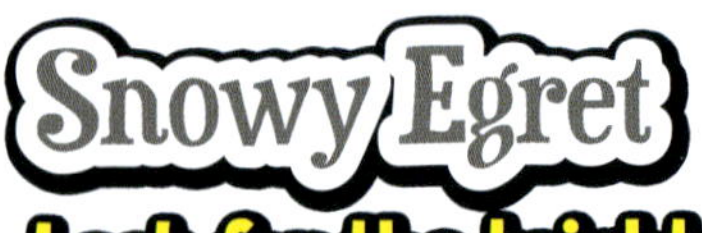

**Look for the bright yellow feet**

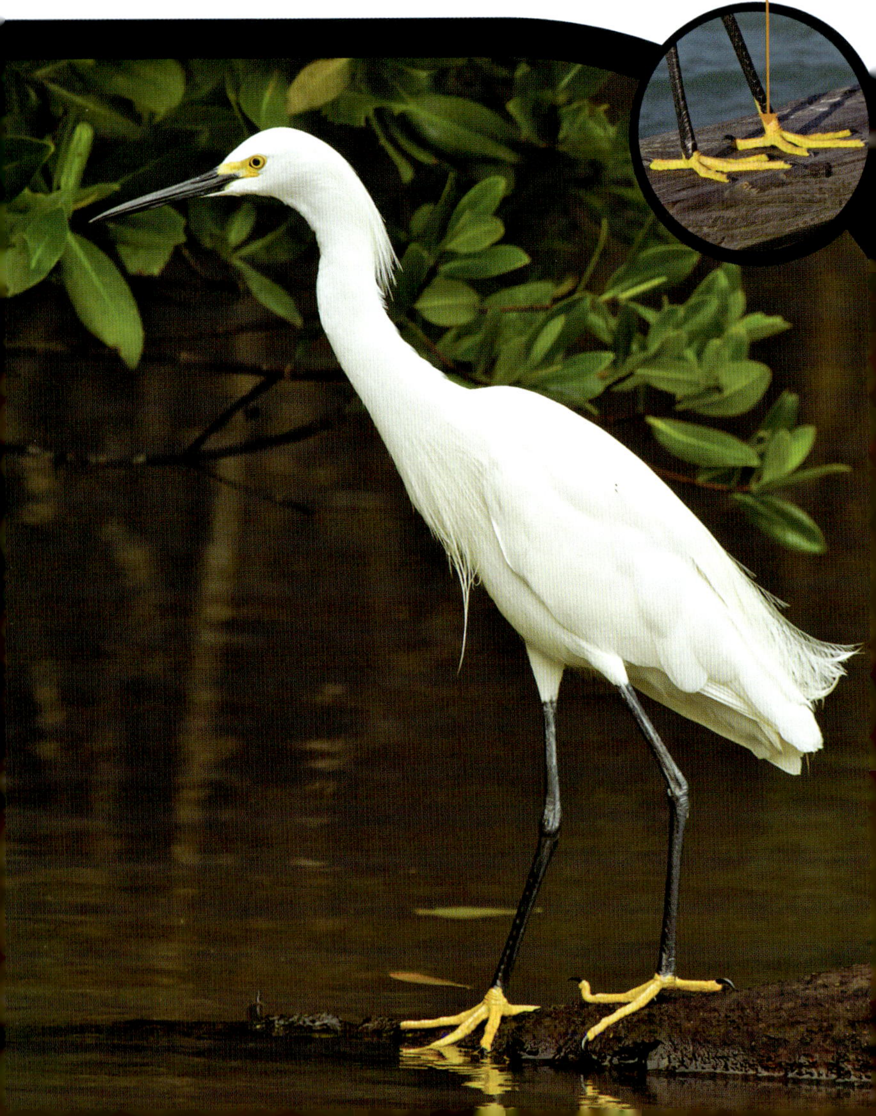

**What to look for:**
all-white egret with a black bill and legs, bright-yellow feet, and long feather plumes on the head, neck, and back; yellow patch at the base of the bill

**Where you'll find them:**
in wetlands and often with other egrets

**Calls and songs:**
usually silent; when startled, gives a loud, raspy, nasal **call** as it flies away

**On the move:**
flies with its neck in an S shape and legs trailing

**What they eat:**
aquatic insects and small fish

**Nest:**
platform; in a **colony** that may have up to several hundred nests; nests are low in shrubs that are 5–10 feet tall or are on the ground, usually mixed among other egret and heron nests

**Eggs, chicks, and childcare:**
3–5 light blue-green eggs; parents alternate sitting on the eggs and feeding the **hatchlings**

**Spends the winter:**
in Florida, the Gulf Coast, southern states, and Mexico

**REAL QUICK**

Size
22–26"

Nest
PLATFORM

Feeder
NONE

summer
migration

SAW IT!

**STAN'S COOL STUFF**

The Snowy Egret was hunted to near extinction in the late 1800s for its long, handsome feather plumes. It hunts actively for **prey**, moving around quickly in the water. It uses its feet to stir up small fish and aquatic insects, which it swiftly snaps up to eat.

# White Ibis

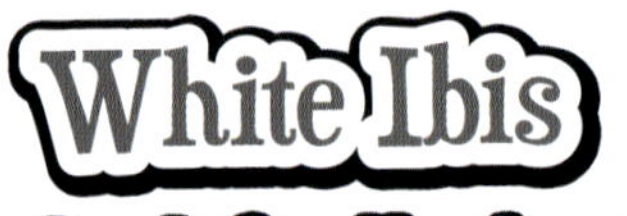

**Look for the long, down-curving bill**

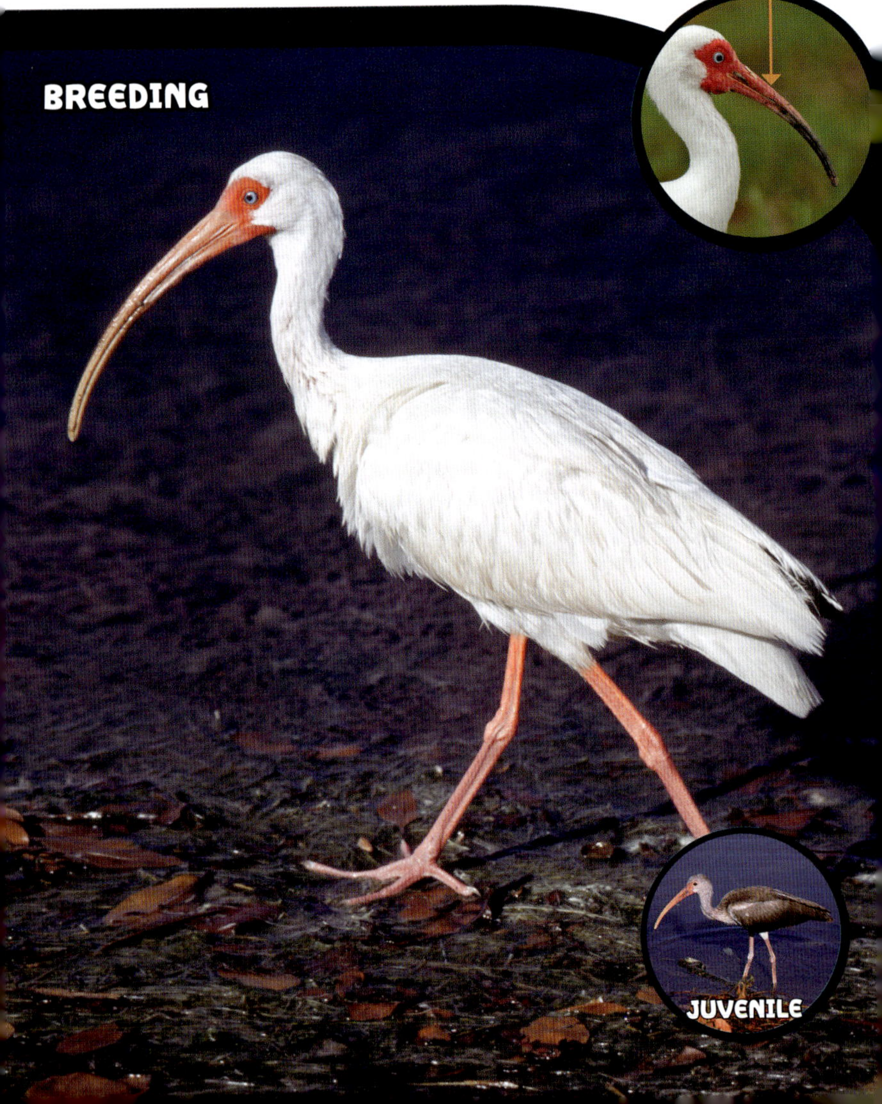
BREEDING
JUVENILE

**What to look for:**
white bird with a long, down-curving orange-to-red bill; pink skin on the face, pink legs, and black wing tips, seen only in flight; juvenile **plumage** is brown and white for the first two years

**Where you'll find them:**
in freshwater and saltwater habitats, but it prefers places with fresh water

**Calls and songs:**
just makes short grunts at the nesting **colony**

**On the move:**
flies in groups of 30 or more birds

**What they eat:**
aquatic bugs, crayfish, and other **crustaceans**, fish

**Nest:**
platform, in a large colony; builds a well-made nest with sticks

**Eggs, chicks, and childcare:**
2–3 light-blue eggs with dark marks; Mom and Dad share **incubation** duty and feed the young

**Spends the winter:**
doesn't **migrate** in Virginia

**REAL QUICK**

Size
23-27"

Nest
PLATFORM

Feeder
NONE

year-round

SAW IT!

## STAN'S COOL STUFF

The White Ibis is a common in many places. Colonies with over 30,000 birds have been reported. Males are larger than females and have longer bills. Juveniles are brown, unlike the adults. The name "ibis" comes from the Latin/Greek word for a group of birds.

# Great Egret

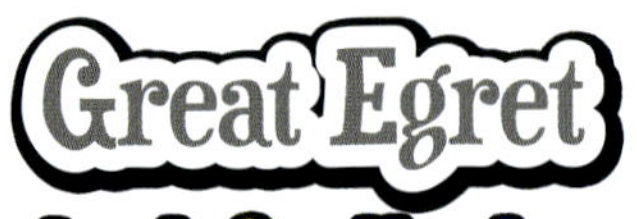

**Look for the long, thin white neck**

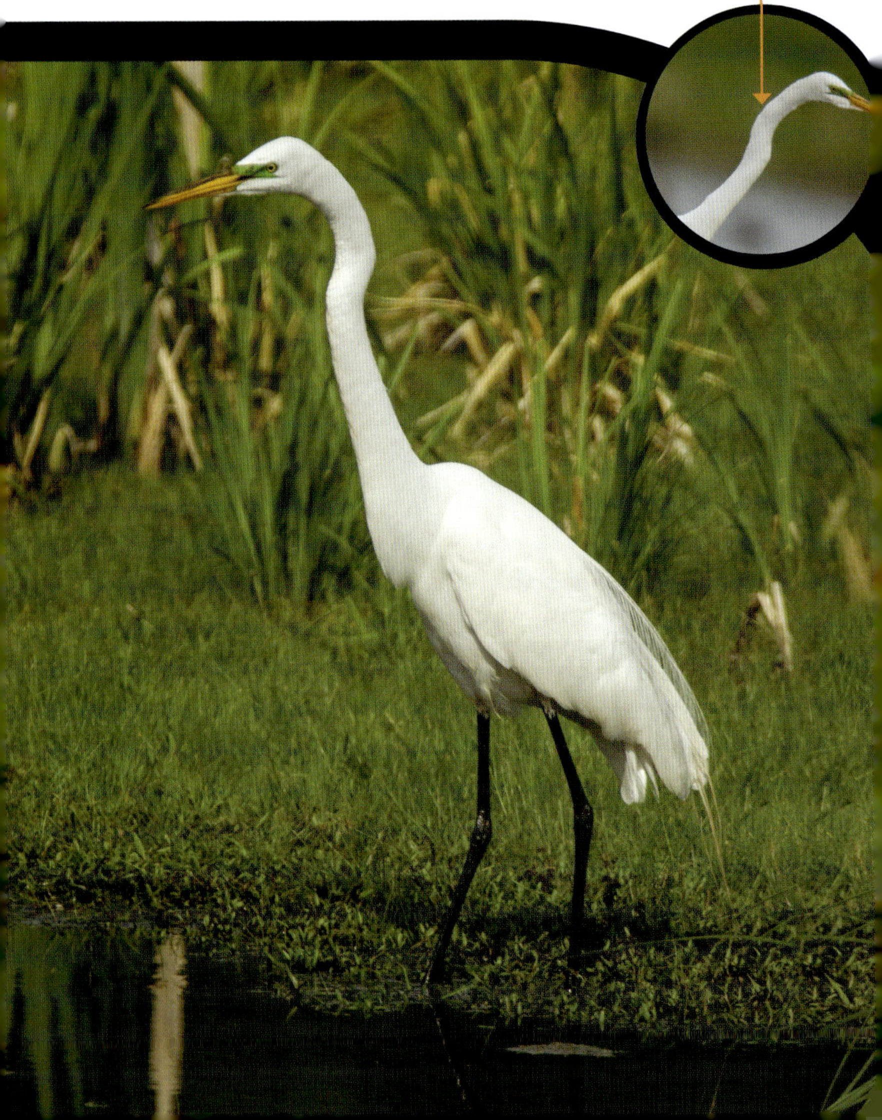

**What to look for:**
tall and thin white egret with a long neck, long legs, and a long, pointed yellow bill

**Where you'll find them:**
shallow wetlands, ponds, and lakes

**Calls and songs:**
gives a loud, dry croak if disturbed or when it squabbles for a nest site at the **colony**

**On the move:**
holds its neck in an S shape during flight; slowly stalks in shallow water, looking for fish to spear with its sharp bill

**What they eat:**
small fish, aquatic insects, frogs, and crayfish

**Nest:**
platform, in a colony of up to 100 birds

**Eggs, chicks, and childcare:**
2–3 light-blue eggs; Mom and Dad sit on the eggs and give food to the **hatchlings**

**Spends the winter:**
some stay in Virginia; southern states

**REAL QUICK**

Size
36–40"

Nest
PLATFORM

Feeder
NONE

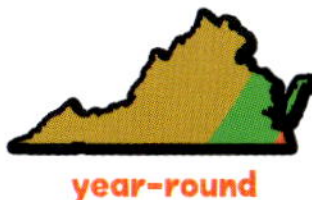

year-round
summer
migration

SAW IT!

**STAN'S COOL STUFF**

From the 1800s to the early 1900s, the Great Egret was hunted to near extinction for its beautiful long plumes, which were used to decorate women's hats. The plumes grow near the tail during the breeding season. Today, the egret is a protected bird.

# Tundra Swan

**Look for small yellow mark in front of eye**

**What to look for:**
large all-white swan with black bill, legs, and feet; small yellow mark in front of each eye

**Where to find them:**
coastal estuaries

**Calls and songs:**
gives a high-pitched, whistle-like call

**On the move:**
flies in large V-shaped wedges; many birds together

**What they eat:**
plants and aquatic insects

**Nest:**
ground; Mom and Dad build it

**Eggs, chicks, and childcare:**
4–5 creamy-white eggs; Mom and Dad take turns sitting on the eggs and feeding the babies

**Spends the winter:**
in eastern Virginia

**REAL QUICK**

Size
**50–54"**

Nest
**GROUND**

Feeder
**NONE**

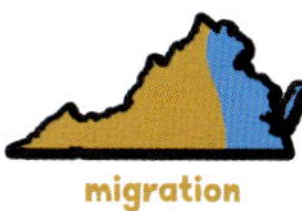

migration
winter

**SAW IT!**

## STAN'S COOL STUFF

This swan nests in the tundra of the Arctic, but it comes to the East Coast (Maryland and the Carolinas) for the winter. It was once called Whistling Swan. Young stay with their parents until they are 1 year old. Always seen in large flocks.

# American Goldfinch

**Look for the black forehead**

MALE

FEMALE

**What to look for:**
bright canary-yellow bird with a black forehead, wings, and tail; female is olive yellow and lacks a black forehead; winter male resembles the female

**Where you'll find them:**
open fields, scrubby areas, woodlands, backyards

**Calls and songs:**
male sings a pleasant high-pitched song; gives **twitter** calls during flight

**On the move:**
appears roller-coaster-like in flight

**What they eat:**
loves seeds and insects; comes to seed (especially thistle) feeders

**Nest:**
cup; builds its nest in late summer and lines the cup with the soft, silky down from wild thistle

**Eggs, chicks, and childcare:**
4–6 pale-blue eggs; Mom incubates the eggs and Dad pitches in to help her feed the babies

**Spends the winter:**
in Virginia and southern states; flocks of up to 20 birds move around in winter to find food

**REAL QUICK**

Size
5"

Nest
CUP

Feeder
TUBE OR HOPPER

year-round

SAW IT!

**STAN'S COOL STUFF**

The American Goldfinch is often called Wild Canary due to its canary-colored **plumage**. This feeder bird is almost always in small flocks, visiting thistle tube feeders that offer Nyjer seed. A late-nesting bird with most nesting in July through mid-September.

# Yellow Warbler

Look for the orange streaks on the chest

MALE

FEMALE

**What to look for:**
yellow bird with thin orange streaks on the chest and belly; female lacks orange streaks

**Where you'll find them:**
gardens and shrubby areas near water, backyards

**Calls and songs:**
male sings a string of sweet notes, sounding like "sweet, sweet, sweet, I'm-so-sweet!"

**On the move:**
zooms around shrubs and shorter trees; begins to migrate south in August, migrating at night in mixed flocks of warblers; males return to claim territories 1–2 weeks before females arrive

**What they eat:**
insects

**Nest:**
cup; Mom builds it

**Eggs, chicks, and childcare:**
4–5 white eggs with brown marks; Mom sits on the eggs; Mom and Dad give food to the kids

**Spends the winter:**
in southern states, Mexico, Central America, and South America

**REAL QUICK**

Size
5"

Nest
CUP

Feeder
NONE

summer

SAW IT!

## STAN'S COOL STUFF

The Yellow Warbler is common and widespread in Virginia. It eats small caterpillars and many other bugs on tree leaves. The male is easier to see higher up in trees than the duller female. He sings loudly, zips off to grab a bug, and then starts singing again.

# Baltimore Oriole

**Look for the gray-brown wings**

**What to look for:**
pale-yellow bird with orange tones and gray-brown wings with white wing bars

**Where you'll find them:**
parks, yards, and forests; in treetops, where it feeds on caterpillars

**Calls and songs:**
a fantastic songster, singing loudly; often heard before it is seen

**On the move:**
often returns to the same area year after year

**What they eat:**
insects, fruit, and **nectar**; comes to nectar, orange-half, and grape jelly feeders

**Nest:**
pendulous; an interesting nest that looks like a sock hanging from an outer branch of a tall tree

**Eggs, chicks, and childcare:**
4–5 bluish eggs with brown marks; Mom sits on the eggs; Mom and Dad do the childcare

**Spends the winter:**
in Mexico, Central and South America

**REAL QUICK**

Size
7–8"

Nest
PENDULOUS

Feeder
NECTAR

summer
migration

SAW IT!

## STAN'S COOL STUFF

Orioles come to feeders that offer sugar water (nectar), orange halves, or grape jelly. Parents bring their young to feeders. Young males look like females for their first 1½ years. They're some of the last birds to arrive in spring and some of the first to leave in fall.

## BIRD FOOD FUN FOR THE FAMILY

If you and your family like to do fun projects together, making your own bird food and bird-feeding items might be just the right ones to try. Chances are good that you already have most of the ingredients at home to make delicious and nutritious treats for your wild bird friends.

You'll be doing these projects in the kitchen, so show your mom or dad the following sections. They're written specifically with the whole family in mind. For example, you may need to check with a parent or guardian for help with such tasks as grocery shopping, stovetop cooking, or food preparation, like cutting up fresh fruit.

### Starter Snacks and Fruit Treats

You can start by offering some food that's already in your kitchen. Peanut butter attracts a lot of birds! Simply use a spatula to smear some on the bark of a nearby tree where you can watch it from a window. Or use a piece of firewood: prop it up or hang it with a rope and slather it with peanut butter—then watch the birds go wild.

To offer treats like raisins, dates, and currants, place them outside in a nonbreakable small bowl with a few holes drilled in the bottom for drainage. Waxwings, robins, and many other birds love small dried fruit, and some will be flying in shortly to get some.

Putting out fresh fruit, such as apples and oranges, is another great way to attract bright and colorful birds to your yard. Cut these into small, manageable pieces, and offer the snacks on the tray of a feeder.

Another cool way to serve an orange is to cut one in half and place the halves sunny-side up on a feeder or branch. This arrangement allows birds to feast easily on the sweet fruit contained inside the rind. Sometimes, it's best to impale the orange half on a nail to stop it from rolling away.

Plain and unsalted nuts, especially peanuts, pecans, and walnuts, make wonderful treats for birds. Simply add these to a feeder tray with seeds or place them in a tube feeder for nuts.

## Easy Bird Food Recipes

Preparing bird food of any kind shows that you care about the birds in your backyard. Now, are you ready to try making some recipes? Below are just a few suggestions. You can find much more online.

### Sweet Homemade Nectar

**Nectar** is a superb food for many birds. Studies of nectar from flowers have shown that the average flower nectar is 25 percent sucrose. Sucrose is a simple sugar, so to make the correct strength of homemade nectar (sugar water), mix a ratio of 1 part sugar to 4 parts water. You'll discover that hummingbirds, orioles, and woodpeckers will thoroughly enjoy the sweet drink that you made.

INGREDIENTS
¼ cup granulated white sugar
1 cup warm water

DIRECTIONS: Add the sugar to the water and stir to dissolve. If you prefer, you can boil the water first so the sugar dissolves more quickly. Cool to room temperature before filling your feeder. Store any extra in the refrigerator or freezer.

NOTES: Never substitute brown sugar or honey for white sugar. Also, there is no need to add red food coloring because the birds will be attracted to any amount of red on any part of your **nectar** feeder.

### Birds-Go-Wild Spread

INGREDIENTS
½ cup raisins
½ cup granola
½ cup oatmeal
½ cup Cheerios
16-ounce jar smooth peanut butter

DIRECTIONS: Mix dry ingredients in a large mixing bowl. Warm the peanut butter in a microwave or place the jar in warm water to soften. Scoop out the softened peanut butter and mix well with dry ingredients until smooth.

Spread on tree bark or smear a few dollops on the tray of a feeder.

### Love-It-Nutty Butter

INGREDIENTS
2 cups shelled peanuts, unsalted
2 cups shelled walnuts, unsalted
¼ cup raisins
3–5 tablespoons coconut oil or other vegetable oil

DIRECTIONS: Toss dry ingredients into a food processor. Start blending. Add oil until the mixture reaches a smooth, thick consistency. Store in refrigerator.

Smear on a wooden board with grooves or spread on tree bark.

## Make Your Own Suet

You and your family can make outstanding **suet** recipes at home, too. Suet is animal fat, often from cows, and there are several convenient ways to get it for your recipes.

A quick way is to purchase plain suet cakes. In store-bought suet, the fat has already been melted down (**rendered**). A cheaper way might be to buy fat trimmings in bulk from your local butcher or large amounts of **lard** at your grocery store. A clever way to get rendered fat from your own kitchen is for an adult to pour fat drippings from cooked bacon, pork, and beef into an empty, clean can. When the fat has cooled and solidified, cover and refrigerate to save for future use.

### Easy-Peasy Suet

INGREDIENTS
1 cup solidified fat of your choice
1 cup chunky peanut butter
3 cups ground cornmeal
1 cup white flour
1 cup black oil sunflower seeds or peanuts

DIRECTIONS: In a large pot, melt the fat over low heat. Do not heat quickly, or the fat might burn. Strain the fat through a **cheesecloth** to remove any chunks, and then pour the liquid back into the pot.

Add the peanut butter to the fat. Stir over low heat until the mixture melts and consistency is smooth. Remove from heat. Add the cornmeal and flour and mix until stiff. Add the sunflower seeds or peanuts, and mix thoroughly.

Pour into a mold or container. With a spatula, spread out the mixture and smooth the top. Cool completely, then cut into squares. Store in freezer.

### Simply Super Suet

INGREDIENTS
2 cups **suet** or **lard**
1 cup peanut butter
2 cups yellow cornmeal
2 cups cracked corn
1 cup black oil sunflower seeds

DIRECTIONS: In a large pot, melt the suet or lard over low heat. Add the peanut butter, stirring until melted and well mixed. Add remaining ingredients, and mix.

Pour into baking pans or forms and allow to cool. Cut into chunks or shapes. Store in freezer.

## Yummy Bird-Feeding Projects

Bird-feeding projects are super activities for families, and they can be a big hit at special occasions, such as birthday parties. These very attractive ornaments and feeders also make unique gifts for the holidays and family celebrations.

### Birdseed Ornaments

INGREDIENTS
cookie cutters in any shape
nonstick cooking spray
½ cup water
3 tablespoons white corn syrup
2½ teaspoons unflavored gelatin
¾ cup white flour
4 cups black oil sunflower seeds
10- to 12-inch pieces of string or **jute** twine

DIRECTIONS: Place the cookie cutters on wax paper and spray with nonstick cooking spray. Set aside.

In a saucepan, bring the water and corn syrup to a boil. Reduce heat and stir in gelatin until completely mixed. Do not overcook.

Transfer the hot liquid to a bowl. Add the flour and mix until smooth. Add the sunflower seeds and mix well. Mixture will now be thick.

Use a spatula to fill each cookie cutter. Be sure to press the seeds into all parts of the shapes. Roll any extra mixture into balls. Poke one hole through each shape and each ball with a pencil or similar object.

When cooled, pop out the ornaments from the cookie cutters. Thread a length of string or twine through each hole and tie the ends to form a loop. Loop each of your ornaments over nearby branches, and watch the birds come to feast!

### Pine Cone Birdseed Feeders

Try your hand at making this fabulous little bird feeder from an ordinary pine cone. It's fun and easy, and everyone in your family can make their own.

INGREDIENTS (per person)
1 pine cone
10- to 12-inch piece of string or **jute** twine
peanut butter
birdseed

DIRECTIONS: Tie a piece of string or twine to a pine cone. Roll the cone in peanut butter, filling the spaces between the "petals" (bracts) and coating the entire surface. Then, roll the cone in birdseed until the seeds completely cover the peanut butter.

Hang the feeder outside where you can see the birds feeding on it, and enjoy the show!

## MORE ACTIVITIES FOR THE BIRD-MINDED

Nothing brings family and friends closer together than a shared interest. Birding and backyard bird feeding are enjoyable, year-round activities that many find appealing. Here are some things to do that are not only fun for everyone but also supportive of the birds.

### Help Birds Build Their Nests

A thoughtful way for the entire family to work together with birds during spring is to put out a variety of soft and flexible natural items to help birds build their nests.

First, gather some everyday materials around your home that birds will use. Here are some excellent items to offer:

- Yarn, cut into 6-inch-long pieces
- Fabric from an old, clean T-shirt, cut into 6-inch-long strips
- **Cotton batting** (used for handicrafts)
- Fuzzy pet hair from a brush

Next, place your materials into an unused, clean **suet** cage. Be sure to let the ends of the yarn and fabric strips hang out, and don't pack the material in tightly. The birds need to be able to take out the items easily.

Hang the cage by a short chain from a tree in early spring, when the birds are starting to construct their nests. And then, wait...

Soon, birds will be flying back and forth to the materials and choosing their favorites. It's a delight to see birds making use of your nesting contributions. Not only have you assisted the bird parents, but you've also helped them provide a comfy home for their families. Good job!

### Make a Bird-watching List

Making a watch list on **poster board** of the birds that have visited your yard is a handicraft project that the whole family will enjoy. You can decorate the poster any way you like, but it's awesome to show pictures of the birds you've spotted and write notes about the sightings.

Each time you see a new species in your yard, mark it on the poster with the date and time of day. Attach it to the refrigerator, or put it in another prominent place where it's easy for everyone in the family to see and add their updates.

Your watch list is also a valuable way to track the arrival of the first hummingbirds and orioles in your area each spring. If you create a new watch list each year, it could reveal trends in the arrival dates. This information would be of interest not only to your family and friends but also to your teachers and local birding organizations.

### Save the Birds with Hawk Cutouts

Another fun and important project is to make hawk cutouts to attach to your windows. These items will help prevent birds from flying into sheets of glass at your home.

In-flight window strikes are one of the major killers of our wild bird friends. Window reflections of the sky, trees, and other natural features in your yard create the illusion for birds that the flight path is clear. When birds see forms of predator birds in the reflections, they will turn away and take another route.

Various web pages show outlines (**silhouettes**) of hawks that you can print and cut out. Check the possibilities, and then pick your favorites.

Tape the cutouts to any large picture windows, as well as other windows and doors with clear glass. This preventive action will greatly reduce the risk of birds crashing headfirst into glass. Then, give yourself a high five for helping to save them.

## Build Your Very Own Birdhouse

A first-rate project for kids and adults to do together is to construct a birdhouse. Building plans are available online for different kinds of birdhouses for different kinds of birds. Give them a once-over, and pick one that you like best for the birds you want nesting nearby.

The instructions online will help you select the right kind of wood and show you how to cut it to the right sizes. Most importantly, the plans will provide the correct size of the entrance hole for the bird, along with how-to instructions for making it. Most birdhouse projects require hand and power tools, so be sure to work with an adult.

You might even want to make multiple birdhouses with your extended family or your neighbors. With everyone doing different tasks, your team can turn out a bluebird box, a wren box, a robin platform, and more!

## Create a Bird-Friendly Yard

There is no better way to support the birds in your area than to plant bird-friendly flowers, bushes, and trees. There are many varieties of these plants, making it easy to choose some that will be ideal for your yard.

Planting perennials that bloom large and showy flowers each year is an outstanding way to feed hummingbirds. Many shrubs produce attractive **nectar**-filled flowers and then, later in the summer, edible fruit, which the birds love. Numerous tree species offer berries and nuts—foods the birds depend on in late fall.

A yard with grass alone just isn't a friendly **habitat** for birds, so sit down with your family and think about putting in a flower garden or adorning your yard with some shrubs and trees. Soon afterward, you'll be hearing the sweet chirping of birds and a rich repertoire of **birdsong** all around you.

## Take a Birding Trip

Everyone loves a good time! For a fun family outing, plan a birding trip to a local park, state park, or national wildlife refuge. In spring, you'll be rewarded with migrating warblers. During summer, all of the nesting birds will be feeding babies. In the fall, **waterfowl** will be superactive. Even in winter, there are many amazing birds to see.

Your local nature center is another good place to see birds. Oftentimes, nature centers have bird feeders set up to attract birds. Stop in after school or early on Saturday mornings to see what comes to the feeders.

### Practice Good Birding

Finding a stray feather or an empty bird nest is exciting when you and your family are sharing time in nature. Examining these wonders and making a sketch or taking photos are always fun educational opportunities. However, everyone should be aware that collecting, possessing, or owning wild bird feathers, nests, and even bird eggs is not permitted under federal law.

It may seem silly that a lost feather or vacant bird nest needs protecting, but very important laws stop people from buying, selling, and trading these items. In the past, a lively market for feathers, bird nests, and also eggs led to the widespread killing of birds, some to near extinction. To prevent this from happening again, strong laws were passed to safeguard all of our bird species.

So enjoy seeing, studying, and learning about birds, but please don't take any feathers, nests, or eggs with you out of their natural environment. Leave them just as you found them, and perhaps someone else will also get the opportunity to benefit from studying them.

## COMMUNITY SCIENCE PROJECTS

I can't think of a more exciting way to learn about birds and expand the birding experience than to take part in a community science project. If you are unfamiliar with community science projects, they are sponsored by organizations in which citizens like yourself can contribute in a meaningful way to actual scientific projects right from your own home! Most projects don't take much time and can be fun family activities, with everyone sharing what they learned about birds.

There are simple community science projects that might have you just count the birds that come to your feeders. Others are more complex and involve more time, effort, and perhaps a little traveling. Either way, I'm sure you can find an enjoyable and educational community project that will be a perfect fit for your family. Give it a try!

Here are some popular projects and resources for you to explore:

The very well-known Christmas Bird Count winter census, FeederWatch and more

birds.cornell.edu/home

Hummingbird migration

journeynorth.org/tm/humm/AboutSpring.html

Finding and counting nesting birds

nestwatch.org

General community science projects for counting birds

birdwatchingdaily.com/featured-stories/year-round-citizen-science-projects

American Kestrel nesting and population study

kestrel.peregrinefund.org

## LEARNING ABOUT BIRDING ON THE INTERNET

Birding online is another fine way to discover more information about birds—plus, it's a terrific way to spend time during rainy summer days and winter evenings after sunset. So check out the websites below, and be sure to share the fabulous things you've learned about birds with your family and friends.

eBird

ebird.org/home

American Birding Association: Young Birders

aba.org/aba-young-birders/

Cornell Lab of Ornithology

birds.cornell.edu/home

Author Stan Tekiela's website

naturesmart.com

In addition, online birding groups can be of valuable assistance to you as well. Facebook has many pages dedicated to specific areas of the state and the birds that live there. These sites are an excellent, real-time resource that will help you spot birds in your region. Consider joining a Facebook birding group.

## GLOSSARY

**birdsong:** A series of musical notes that a bird strings together in a pleasing melody. Also called a song. See *warble*.

**brood:** A group of bird brothers and sisters that hatched at around the same time.

**brood parasites:** Birds that don't nest, incubate or raise families, such as Brown-headed Cowbirds (pg. 21). See *host*.

**call:** A nonmusical sound, often a single note, repeated. See *note*.

**carrion:** A dead and often rotting animal's body, or carcass, that is an important food for many other animals, including birds.

**cheesecloth:** A loosely woven cotton cloth used primarily to wrap cheese but also used to strain particles from liquids.

**colony:** A group of birds nesting together in the same area. The size of a colony can range from two pairs to hundreds of birds.

**coniferous:** A tree or shrub that has evergreen, needle-like leaves and that produces cones.

**cotton batting:** A light, soft cotton material, often used to stuff quilts.

**cover:** A dense area of trees or shrubs where birds nest or hide.

**crop-milk:** A liquid that pigeons and doves regurgitate (spit up) to feed their young.

**crustaceans:** A large, mainly aquatic group of critters, such as crayfish, crabs, and shrimp.

**deciduous:** A tree or shrub that sheds its leaves every year.

**display:** An attention-getting behavior of birds to impress and attract a mate, or to draw predators away from the nest. A display may include dramatic movements in flight or on the ground.

**double-scratch:** A hunting behavior of some ground feeders, such as juncos, involving hopping forward and scratching the ground back with both feet at the same time to uncover seeds.

**egg dumping:** A nesting behavior of some birds, such as Wood Ducks (pg. 129), in which the female lays some of her eggs in another female's nest in addition to her own nest.

**epaulets:** Decorative color patches on the shoulders of a bird, as seen in male Red-winged Blackbirds (pg. 25).

**excavate:** To dig or remove wood or dirt, creating a cavity, hole or tunnel.

**fledge:** The process of developing flight feathers and leaving the nest.

**fledglings:** Young birds that have recently left the nest. See *hatchlings.*

**flock:** A group of the same bird species or a gathering of mixed species of birds. Flocks range from a pair of birds to upwards of 10,000 individuals.

**habitat:** The natural home or environment of a bird.

**hatchlings:** Baby birds that have recently emerged from their eggs. See *fledglings.*

**hood:** The markings on the head of a bird, resembling a hood.

**horns:** A tuft or collection of feathers, usually on top of a bird's head, resembling horns.

**host:** A bird species, such as the Red-winged Blackbird (pg. 25), that takes care of the eggs and babies of other bird species. See *brood parasites*.

**incubation:** The process of sitting on bird eggs in the nest to keep them warm until they hatch.

**iridescent:** A luminous, or bright quality of feathers, with colors seeming to change when viewed from different angles.

**jute:** A string of rough fibers made from plants.

**lard:** Fat from mammals, such as cows and pigs.

**lichen:** A unique partnership of plant and fungi growing together and looking and acting as one organism.

**lores:** The areas on each side of a bird's face between the eye and the base of the bill.

**migrate:** The regular and predictable pattern of seasonal movement by some birds from one region to another, especially to escape winter.

**molt:** The process of dropping old, worn-out feathers and replacing them with new feathers, usually only one feather at a time.

**mute:** The inability to make or produce audible sounds. The Turkey Vulture (pg. 41), for example, is mostly mute.

**nape:** The back of a bird's neck.

**necklace:** The markings around the neck of a bird, as seen in the Killdeer (pg. 115).

**nectar:** A sugar and water solution usually consisting of approximately 25 percent sucrose and 75 percent water, usually found in flowers.

**nestlings:** Young birds that have not yet left the nest. See *hatchlings*.

**note:** A single sound of a call, such as the "chip" note of a Northern Cardinal (pg. 199), that doesn't change pitch or frequency. See *birdsong*.

**pair bond:** The relationship between a male and female bird during the mating season.

**plumage:** The collective set of feathers on a bird at any given time.

**poster board:** A stiff cardboard used for displaying information.

**prey:** Any critter that is hunted and killed by another for food.

**raptor:** A flesh-eating bird of prey that hunts and kills for food. Hawks, eagles, Ospreys, falcons, owls, and vultures are raptors. See *prey*.

**regurgitate:** The process of bringing swallowed food up again to the mouth to feed young birds.

**rendered:** Animal fat that has been reduced or melted down by heating in order to make it pure.

**silhouettes:** Dark shapes or outlines against a lighter background.

**speculum:** A patch of bright feathers on some birds, such as ducks, found on the wings.

**squab:** A young pigeon or dove, usually still in the nest.

**suet:** Animal fat, usually beef, that has been heated and made into cakes to feed birds. See *rendered*.

**trachea:** A large tubelike organ that allows air to pass between the lung and the mouth of a bird. Also called a windpipe.

**tree sap:** The watery liquid that moves up and down within the circulatory system of a tree, carrying nutrients throughout.

**trill:** A fluttering or repeated series of similar-sounding musical notes given by some birds, especially Dark-eyed Juncos (pg. 155). See *note*.

**twitter:** A high-pitched call of a bird. See *call*.

**ultraviolet light:** A kind of light that is visible to birds and insects but unseen by people.

**vegetation:** Any plants, especially those found in a particular habitat.

**warble:** A series of pleasing musical notes strung together and often changing. See *trill*.

**waterfowl:** A group of similar birds with a strong connection to water. Ducks, geese, swans, and others, including Blue-winged Teals (pg. 125), are waterfowl.

## CHECKLIST/INDEX BY SPECIES

Use the circles to checkmark the birds you've seen.

## ABOUT THE AUTHOR

Naturalist, wildlife photographer, and writer Stan Tekiela is the originator of the popular state-specific field guide series that includes *Birds of Virginia Field Guide*. He has authored more than 195 field guides, nature books, children's books, wildlife audio CDs, and playing cards, presenting many species of birds, mammals, reptiles, amphibians, trees, wildflowers, and cacti in the United States.

With a Bachelor of Science degree in Natural History from the University of Minnesota, and as an active professional naturalist for more than 30 years, Stan studies and photographs wildlife throughout the United States and Canada. He has received various national and regional awards for his books and photographs. Also a well-known columnist and radio personality, his syndicated column appears in more than 25 newspapers, and his wildlife programs are broadcast on a number of Midwest radio stations. Stan can be contacted via his website, naturesmart.com.